AF614696

METHODS IN MOLECULAR BIOLOGY™

For other titles published in this series, go to
www.springer.com/series/7651

Mammalian Cell Viability

Methods and Protocols

Edited by

Martin J. Stoddart

AO Research Institute Davos, Davos Platz, Switzerland

Editor
Martin J. Stoddart, Ph.D.
AO Research Institute Davos
Davos Platz, Switzerland
martin.stoddart@aofoundation.org

ISSN 1064-3745 e-ISSN 1940-6029
ISBN 978-1-61779-107-9 e-ISBN 978-1-61779-108-6
DOI 10.1007/978-1-61779-108-6
Springer New York Dordrecht Heidelberg London

Library of Congress Control Number: 2011925550

Printed on acid-free paper

Humana Press is part of Springer Science+Business Media (www.springer.com)

Preface

Viability of cells is one of the most fundamental measurements made during studies in cell biology. Whether the question is one of basic cell survival, or whether it is being used to correlate cell number to some other factors such as matrix synthesis, an estimate of cell viability is universally required. The specific method used will greatly influence the interpretation of the data. While many viability methods have been used for decades, there have been recent developments which offer increased sensitivity, throughput, and specificity. The particular type of cell death, apoptotic or necrotic, is becoming increasingly important. This requires multiplexing of methods, or methods that are able to distinguish between the different cell states. This book aims to bring together a wide array of methods in order to assist the reader to determine which is most suitable for them. Certain methods will provide information about the population as a whole, while others determine viability on single cell level. In some cases, it may also be important to realize that the method, while producing an answer, is not suitable for the application applied. It is hoped that the pros and cons of each method will become clear. Many methods have been devised for monolayer cell culture, and although they can often be translated into a three-dimensional system, care must be taken to limit artifacts.

Multiplexing assays is one mechanism by which many sets of data can be obtained from a small number of samples. Owing to the wide array of various viability assays, with the option for both colorimetric and fluorescent measurements, the potential combinations are endless and could not be covered within a book. While many chapters within this book employ more than one assay, two chapters specifically dealing with 96-well multiplexing are provided in order to illustrate practical examples. These can also form a basis which can be modified and utilized with other assays. A number of kits are now commercially available for both single assays and multiplexing, and often the underlying technology of the different kits used is comparable. It is for this reason a good understanding of the reagent used, and how it functions, is crucial when accurately interpreting the data obtained.

This book describes methods from the most basic level, which can be performed in any laboratory, to more complex methods which require specialist pieces of equipment. Initially, the chapters are focused on methods for monolayer and suspension cells; later chapters describe methods for determining viability within tissue sections and three-dimensional culture systems. Finally, methods requiring highly specialized equipment are described in order to explain what is possible. The last chapter aims to provide some guidance as to how automated image analysis can reduce time and inconsistency of quantifying large numbers of images of live or dead cells.

In preparing *Mammalian Cell Viability: Methods and Protocols*, the aim has been to produce a self-contained laboratory manual which is useful for both experienced researchers and those new to the field. I hope that everyone can learn something from this book. Finally, I wish to thank all the contributing authors, as well as John Walker and the staff at Humana Press for seeing this project through.

Davos Platz, Switzerland ***Martin J. Stoddart***

Contents

Contributors

YAMA A. ABASSI • *ACEA Biosciences Inc., San Diego, CA, USA*
JEAN-MARIE AERTS • *Division M3-BIORES: Measure, Model & Manage Bioresponses, Katholieke Universiteit Leuven, Heverlee, Belgium*
ALEXIS D. ARMOUR • *Department of Surgery, University of Missouri, Columbia, MO, USA*
FRANCK A. ATIENZAR • *Investigative Non-Clinical Safety, Non-Clinical Development, UCB Pharma SA, Braine-l'Alleud, Belgium*
SEBASTIAN DE BOODT • *Division M3-BIORES: Measure, Model & Manage Bioresponses, Katholieke Universiteit Leuven, Heverlee, Belgium*
STEVEN T. BOYCE • *Department of Surgery, University of Cincinnati, Cincinnati, OH, USA; Department of Biomedical Engineering, University of Cincinnati, Cincinnati, OH, USA; Research Department, Shriners Hospitals for Children, Cincinnati, OH, USA*
OLIVIER BRAISSANT • *Laboratory of Biomechanics & Biocalorimetry, Coalition for Clinical Morphology & Biomedical Engineering, University of Basel, Basel, Switzerland*
CLAUDIO BRANCOLINI • *Dipartimento di Scienze e Tecnologie Biomediche, Università degli Studi di Udine, Udine, Italy*
ADELINE CAMBON-BINDER • *Laboratoire de Bio-ingéniérie et Biomécanique Ostéo-articulaires, UMR, CNRS, 7052, Paris, France*
SAMANTHA CHAN • *ARTORG, Center for Biomedical Engineering Research, Institute for Surgical Technology and Biomechanics, University of Bern, Bern, Switzerland*
EWA M. CZEKANSKA • *AO Research Institute, Davos Platz, Switzerland*
ALMA U. "DAN" DANIELS • *Laboratory of Biomechanics & Biocalorimetry, Coalition for Clinical Morphology & Biomedical Engineering, University of Basel, Basel, Switzerland*
ZBIGNIEW DARZYNKIEWICZ • *Brander Cancer Research Institute, New York Medical College, Valhalla, NY, USA*
STÉPHANE DHALLUIN • *Investigative Non-Clinical Safety, Non-Clinical Development, UCB Pharma SA, Braine-l'Alleud, Belgium*
BENJAMIN GANTENBEIN-RITTER • *ARTORG, Center for Biomedical Engineering Research, Institute for Surgical Technology and Biomechanics, University of Bern, Bern, Switzerland*
HELGA H.J. GERETS • *Investigative Non-Clinical Safety, Non-Clinical Development, UCB Pharma SA, Braine-l'Alleud, Belgium*
SIBYLLE GRAD • *AO Research Institute, Davos Platz, Switzerland*
SVENJA ILLIEN-JÜNGER • *AO Research Institute, Davos Platz, Switzerland; Department of Orthopaedics, Mount Sinai Medical Centre, New York, USA*
KATHARINA JÄHN • *School of Dentistry Department of Oral Biology University Missouri, Kansas City, Missouri, USA*

Won-Kyu Ju • *Department of Ophthalmology, Hamilton Glaucoma Center, University of California San Diego, La Jolla, CA, USA*
Ning Ke • *ACEA Biosciences Inc., San Diego, CA, USA*
Keun-Young Kim • *Center for Research on Biological Systems, National Center for Microscopy and Imaging Research and Department of Neuroscience, University of California San Diego, La Jolla, CA, USA*
László Kupcsik • *Genetics Unit, Shriners Hospital for Children, Montreal, QC, Canada*
Yang V. Li • *Department of Biomedical Sciences, College of Osteopathic Medicine, Ohio University, Athens, OH, USA*
Delphine Logeart-Avramoglou • *Laboratoire de Bio-ingéniérie et Biomécanique Ostéo-articulaires, UMR, CNRS, 7052 Paris, France*
Kristine S. Louis • *University Health Network, MaRS Centre, Toronto Medical, Toronto, ON, Canada*
Stewart Martin • *Translational Radiation Biology Research Group, Department of Clinical Oncology, University of Nottingham, Nottingham University Hospitals NHS Trust, Nottingham, UK*
Norbert Meyer • *Schweizerisches Institut für Allergie und Asthma Forschung (SIAF), Davos, Switzerland*
Richard A. Moravec • *Promega Corporation, Madison, WI, USA*
Andrew L. Niles • *Promega Corporation, Madison, WI, USA*
Ioan Notingher • *School of Physics and Astronomy, The University of Nottingham, Nottingham, UK*
Karim Oudina • *Laboratoire de Bio-ingéniérie et Biomécanique Ostéo-articulaires, UMR, CNRS, 7052, Paris, France*
Gabriela Paroni • *Department of Biochemistry and Molecular Biology, Mario Negri Institute for Pharmacological Research, Milan, Italy*
Heather M. Powell • *Department of Materials Science and Engineering, The Ohio State University, Columbus, OH, USA; Department of Biomedical Engineering, The Ohio State University, Columbus, OH, USA*
Terry L. Riss • *Promega Corporation, Madison, WI, USA*
Andre C. Siegel • *University Health Network, MaRS Centre, Toronto Medical, Toronto, ON, Canada*
Joanna Skommer • *School of Biological Sciences, University of Auckland, Auckland 1142, New Zealand*
Jos Vander Sloten • *Division of Biomechanics and Engineering Design, Katholieke Universiteit Leuven, Heverlee, Belgium*
Pieter Spaepen • *Katholieke Hogeschool Kempen, Geel, Belgium*
Christoph M. Sprecher • *AO Research Institute, Davos Platz, Switzerland*
Martin J. Stoddart • *AO Research Institute Davos, Davos Platz, Switzerland*
Christian J. Stork • *Department of Biomedical Sciences, College of Osteopathic Medicine, Ohio University, Athens, OH, USA*
Sophie Verrier • *AO Research Institute, Davos Platz, Switzerland*
Xiaobo Wang • *ACEA Biosciences Inc., San Diego, CA, USA*

DONALD WLODKOWIC • *The BioMEMS Research Group, Department of Chemistry, University of Auckland, Auckland 1142, New Zealand*
CAROLINE WOOLSTON • *Translational Radiation Biology Research Group, Department of Clinical Oncology, University of Nottingham, Nottingham University Hospitals NHS Trust, Nottingham, UK*
XIAO XU • *ACEA Biosciences Inc., San Diego, CA, USA*
MAYA ZIMMERMANN • *Schweizerisches Institut für Allergie und Asthma Forschung (SIAF), Davos, Switzerland*
ALINA ZOLADEK • *School of Physics and Astronomy, The University of Nottingham, Nottingham, UK*

Chapter 1

Cell Viability Assays: Introduction

Martin J. Stoddart

Abstract

The measurement of cell viability plays a fundamental role in all forms of cell culture. Sometimes it is the main purpose of the experiment, such as in toxicity assays. Alternatively, cell viability can be used to correlate cell behaviour to cell number, providing a more accurate picture of, for example, anabolic activity. There are wide arrays of cell viability methods which range from the most routine trypan blue dye exclusion assay to highly complex analysis of individual cells, such as using RAMAN microscopy. The cost, speed, and complexity of equipment required will all play a role in determining the assay used. This chapter aims to provide an overview of many of the assays available today.

Key words: Apoptosis, Viability, Multiplexing, 3D, 2D

Cell viability is often defined as the number of healthy cells in a sample. Often the same assays used to determine viability are used repeatedly over a period of time to investigate cell proliferation within a population. Cell viability methods loosely can be categorised into those which analyse whole populations and those which involve analysis of individual cells. Generally speaking, the population analysis is more rapid, but gives a less detailed result than those which involve viability measurements on the single cell level. One of the earliest methods for assessing cell viability was trypan blue dye exclusion assay, which is still widely used today. It is based on the principle that viable cells have an intact cell membrane which can therefore exclude the trypan blue dye. Dead cells take up trypan blue, and appear blue as a consequence, as their membrane is no longer able to control the passage of macromolecules. The assay requires the cells to be in a single cell suspension and they are then visualised and counted under a microscope using a haemocytometer of a defined volume or using automated counting devices which have recently become available. From these counts, it is relatively simple to calculate the total number of cells and the percent of viable cells within a population.

Martin J. Stoddart (ed.), *Mammalian Cell Viability: Methods and Protocols*, Methods in Molecular Biology, vol. 740,
DOI 10.1007/978-1-61779-108-6_1, © Springer Science+Business Media, LLC 2011

Over time the assays for viability have become more complex. Increasingly, dyes which rely on the metabolic activity of cells are gaining favour as they can be performed on adherent cells and therefore lend themselves to high-throughput analysis. The final measurement is also carried out using a plate reader which when screening many treatment groups provides a major advantage. When determining which assay to use, preliminary studies should be carried out to determine which is the optimal assay and when should it be applied (1).

The first, and probably best known, metabolic dye is 3-(4,5-dimethylthiazol-2-yl)-2,5-diphenyltetrazolium bromide (MTT) and the development of a 96-well assay format dramatically increased the capacity (2). This assay relies on the conversion of soluble tetrazolium into insoluble blue formazan crystals by reduction. While it has generally been believed that this conversion takes place in the mitochondria, there is increasing evidence that this may not necessarily be the case (3). Other similar dyes have become increasingly popular, such as XTT (sodium 3′-[1-(phenylaminocarbonyl)-3,4-tetrazolium)-bis-(4-methoxy-6-nitro) benzene sulphonic acid hydrate) (4), MTT (3-(4,5-dimethylthiazol-2-yl)-2,5-diphenyltetrazolium bromide), MTS (3-(4,5-dimethylthiazol-2-yl)-5-(3-carboxymethoxyphenyl)-2-(4-sulphophenyl)-2H-tetrazolium), and WST derivatives. XTT and MTS are used in the presence of phenazine methosulphate (PMS) which increases their sensitivity (4, 5). These newer derivatives are reportedly more sensitive but their major advantage is that they have fewer steps as the converted product is released into the medium, meaning no step is required to dissolve the insoluble product (a required step when using MTT). However, MTT is very robust and is metabolised by most cell types, whereas some of the newer alternatives are not suitable for all cells.

Resazurin-based methods (Alamar Blue) are also routinely used as they provide the advantage that it can be measured by either colourimetry or fluorimetry. Greater sensitivity is achieved using the fluorescent property where as few as 80 cells can give a reproducible and sensitive signal (6). However, care must be taken when measuring fluorescence at higher cell density to avoid over-reduction of Alamar Blue into the colourless and non-fluorescent hydroresorufin (6). All assays based on reduction must be tested without any cells in the medium to determine any cross reactivity with the compound to be tested.

Live/dead viability staining describes a number of potential dyes where one specifically stains live cells (usually resulting in green fluorescence) while the other dye stains dead cells (usually with red fluorescence). An example would be calcein AM with ethidium homodimer-1 but there are many combinations which generally involve a membrane-permeable dye which is metabolised within viable cells, mixed with a membrane-impermeable DNA binding molecule. The calcein AM is membrane permeable and is cleaved by esterases in live cells to yield cytoplasmic green fluorescence. The membrane-impermeable ethidium homodimer-1 labels nucleic acids of membrane-compromised cells (i.e. dead) with red fluorescence. The ratio of live to dead then can easily be determined by simple counting. In assays such as live/dead staining, there is also the possibility to include macro-based analysis within the workflow. A carefully written macro in combination with image analysis can provide rapid quantification of hundreds of images and yet this method is currently underutilised by many laboratories.

Multiplexing assays is one mechanism by which many sets of data can be obtained from a small number of samples. Due to the wide array of various viability assays, with the option for both colorimetric and fluorescent measurements the potential combinations are endless and could not be covered within one book. While a number of the chapters within this book employ more than one assay, two chapters specifically dealing with 96-well multiplexing are provided in order to illustrate practical examples. These can also form a basis which can be modified and utilised with other assays.

A number of kits are now commercially available for both single assays and multiplexing, and often the underlying technology the different kits used are comparable. It is for this reason a good understanding of the reagent used, and how it functions, is crucial when accurately interpreting the data obtained.

While some earlier assays such as chromium release assay (7) relied on the use of radioactive isotopes, safety and environmental issues have led to a sharp decline in their use, for this reason they are not covered within this book.

The mechanism of cell death is also increasingly being investigated with the need to determine whether death was as a result of apoptosis or necrosis. The term apoptosis was first proposed in 1972 (8) and is a programmed sequence of events that is responsible for removing unwanted cells during normal development. Apoptosis also occurs in the adult where it is responsible for removing cells at the end of their useful lifespan. Apoptosis can also occur when a cell is damaged beyond repair, infected with a virus, or undergoing a period of stress. Necrosis is an alternative mechanism of cell death which is generally considered more traumatic. It occurs due to infection, toxins, or trauma. There are a number of mechanisms to distinguish the type of cell death which is occurring, both on single cells and on populations. Caspase activity can be investigated in a number of ways and as such has a chapter dedicated only to this method of investigating apoptosis. During the apoptotic pathway, caspases are activated and cleave a number of cell proteins as part of the programmed cell death pathway (for review see ref. 9). This activity can be detected in a number of ways in order to determine the level of apoptosis present within a sample. A number of companies have now also developed ELISA-based methods for quantifying apoptosis.

Alternatively, using a combination of DNA binding compounds such as SYTO probes (10) and propidium iodide it is possible to determine the type of cell death by investigating the cell staining pattern. During cell death changes occur in the integrity of the cell membrane. Healthy and early apoptotic cells have an intact cell membrane which is impermeable to DNA binding dyes such as propidium iodide (PI) and 7-aminoactinomycin D (7-AAD) (11). Staining cells with these dyes can give information as to their membrane integrity. This can be done on a single cell level by using a flow cytometer, and quantification of the individual populations can be rapidly achieved. Changes in membrane properties, which occur during apoptosis, can also be used as a method to determine whether a particular cell is engaged in programmed cell death. Healthy cells have phosphatidylserines on the inner leaflet of the plasma membrane. During early apoptosis, the phosphatidylserines flip onto the outer leaflet of the plasma membrane where they can be easily stained by labelled annexin V (12). This feature, again in combination with DNA binding dyes which are impenetrable to intact membranes, can be used to determine the apoptotic state of a cell.

Increasingly, investigators would like to determine cell viability within 3D tissues. This adds a greater level of complexity than standard 2D culture. It is critical to ensure that the assay used is able to penetrate the tissue and that rate-limiting mass transfer is not an issue. The usual way to confirm this is a day 0 control where the viability is known to be high. We have previously shown that methods routinely used for monolayer culture are not always suitable in 3D constructs (13). When investigating the viability of osteocytes within whole cancellous bone cores, MTT and live/dead staining led to a ring of viable cells with the central areas appearing to be dead, even though the bone core had been freshly prepared. This could be overcome by cooling the bone core and staining solution to 4°C and allowing the reagent to penetrate for 1–2 h. Transferring the core to 37°C then led to the expected viability staining (13) (see Fig. 1).

Alternatively, staining can be used on sections that enable investigations into viability of tissues. The terminal transferase-mediated dUTP nick end-labelling (TUNEL) assay is commonly used to determine apoptosis in whole tissue sections (14). It is based on the fact that apoptosis leads to a controlled digestion of the nuclear DNA which results in the presence of nicks in the DNA. These can be identified by terminal deoxynucleotidyl transferase, an enzyme that catalyses the addition of dUTPs that are secondarily labelled with a marker. The regulated degradation of DNA also leads to the formation of a distinct banding pattern which can be readily distinguished using agarose gel electrophoresis (15).

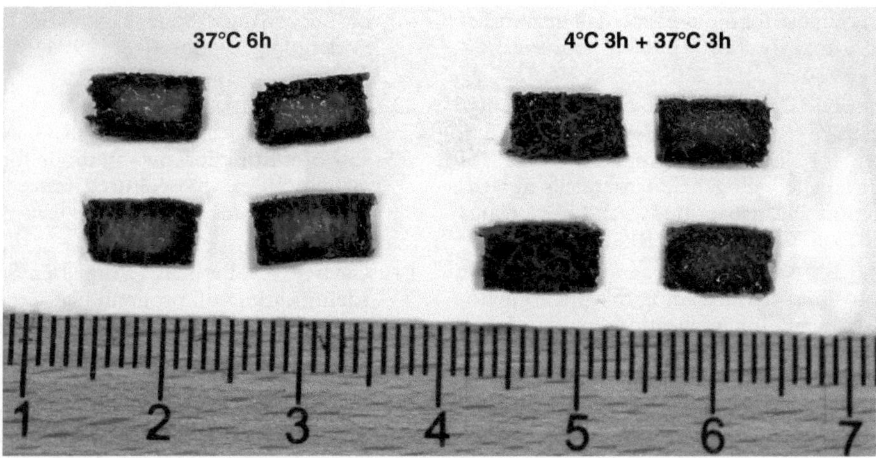

Fig. 1. The viability of whole cancellous bone cores was investigated using MTT. Freshly prepared cores were incubated for either 6 h at 37°C or for 3 h at 4°C followed by 3 h at 37°C. As can be seen by the lack of substrate conversion at the centre of the 6 h at 37°C cores, the reagent was not able to penetrate to the centre of the tissue under these conditions. This indicates that under conditions of low diffusion rates, the cells at the outer regions of the tissue can metabolise the reagent faster than the reagent can penetrate. This shows care must be taken when using metabolic labels in 3D samples.

Release of lactate dehydrogenase (LDH) into cell culture medium is often used to assess cell death. As the cell membrane integrity breaks down, the enzyme is released and can then be quantified. However, the stability of LDH within cells can also be used to determine cell viability in whole sections of fresh tissue (16). As LDH is stable for 36–48 h after cell death (17), tissue sections can be prepared and the cell viability analysed even though the cell does not need to be viable at the time of assay.

In conclusion, the optimal viability assay depends on the cell type, culture conditions applied, and the specific questions being asked. When detailed information is required it may be more suitable to apply more than one method in order to confirm the results of individual assays.

References

1. Riss T.L., and Moravec R.A. Use of multiple assay endpoints to investigate the effects of incubation time, dose of toxin, and plating density in cell-based cytotoxicity assays. (2004) *Assay. Drug Dev. Technol.* **2**, 51–62.
2. Mosmann T. Rapid colorimetric assay for cellular growth and survival: application to proliferation and cytotoxicity assays. (1983) *J. Immunol. Methods* **65**, 55–63.
3. Liu Y., Peterson D.A., Kimura H., and Schubert D. Mechanism of cellular 3-(4,5-dimethylthiazol-2-yl)-2,5-diphenyltetrazolium bromide (MTT) reduction. (1997) *J Neurochem* **69**, 581–593.
4. Scudiero D.A., Shoemaker R.H., Paull K.D., Monks A., Tierney S., Nofziger T.H., Currens M.J., Seniff D., and Boyd M.R. Evaluation of a soluble tetrazolium/formazan assay for cell growth and drug sensitivity in culture using human and other tumor cell lines. (1988) *Cancer Res.* **48**, 4827–4833.
5. Cory A.H., Owen T.C., Barltrop J.A., and Cory J.G. Use of an aqueous soluble tetrazolium/formazan assay for cell growth assays in culture. (1991) *Cancer Commun.* **3**, 207–212.
6. O'Brien J., Wilson I., Orton T., and Pognan F. Investigation of the Alamar Blue (resazurin)

fluorescent dye for the assessment of mammalian cell cytotoxicity. (2000) *Eur J Biochem* **267**, 5421–5426.

7. Brunner K.T., Mauel J., Cerottini J.C., and Chapuis B. Quantitative assay of the lytic action of immune lymphoid cells on 51-Cr-labelled allogeneic target cells in vitro; inhibition by isoantibody and by drugs. (1968) *Immunology* **14**, 181–196.
8. Kerr J.F., Wyllie A.H., and Currie A.R. Apoptosis: a basic biological phenomenon with wide-ranging implications in tissue kinetics. (1972) *Br. J. Cancer* 26, 239–257.
9. Kurokawa M., and Kornbluth S. Caspases and kinases in a death grip. (2009) *Cell* **138**, 838–854.
10. WlodkowicD.,SkommerJ.,andDarzynkiewicz Z. SYTO probes in the cytometry of tumor cell death. (2008) *Cytometry A* **73**, 496–507.
11. Schmid I., Krall W.J., Uittenbogaart C.H., Braun J., and Giorgi J.V. Dead cell discrimination with 7-amino-actinomycin D in combination with dual color immunofluorescence in single laser flow cytometry. (1992) *Cytometry* **13**, 204–208.
12. Koopman G., Reutelingsperger C.P., Kuijten G.A., Keehnen R.M., Pals S.T., and van Oers M.H. Annexin V for flow cytometric detection of phosphatidylserine expression on B cells undergoing apoptosis. (1994) *Blood* **84**, 1415–1420.
13. Stoddart M.J., Furlong P.I., Simpson A., Davies C.M., and Richards R.G. A comparison of non-radioactive methods for assessing viability in ex vivo cultured cancellous bone: technical note. (2006) *Eur. Cell Mater.* **12**, 16–25.
14. Gavrieli Y., Sherman Y., and Ben Sasson S.A. Identification of programmed cell death in situ via specific labeling of nuclear DNA fragmentation. (1992) *J. Cell Biol.* **119**, 493–501.
15. Wyllie A.H. Glucocorticoid-induced thymocyte apoptosis is associated with endogenous endonuclease activation. (1980) *Nature* **284**, 555–556.
16. Wong S.Y., Dunstan C.R., Evans R.A., and Hills E. The determination of bone viability: a histochemical method for identification of lactate dehydrogenase activity in osteocytes in fresh calcified and decalcified sections of human bone. (1982) *Pathology* **14**, 439–442.
17. Wong S.Y., Kariks J., Evans R.A., Dunstan C.R., and Hills E. The effect of age on bone composition and viability in the femoral head. (1985) *J Bone Joint Surg Am* **67**, 274–283.

Chapter 2

Cell Viability Analysis Using Trypan Blue: Manual and Automated Methods

Kristine S. Louis and Andre C. Siegel

Abstract

One of the traditional methods of cell viability analysis is the use of trypan blue dye exclusion staining. This technique has been the standard methodology used in academic research laboratories and industrial biotechnology plants. Cells were routinely counted manually with a hemocytometer. In recent years, modern automated instrumentation has been introduced to supplement this traditional technique with the efficiency and reproducibility of computer control, advanced imaging, and automated sample handling.

Key words: Cell viability, Cell counting, Trypan blue, Hemocytometer, Automation

1. Introduction

Dye exclusion methods are traditionally used to assess cell viability, with trypan blue being one of the most common. Trypan blue is a vital stain that leaves nonviable cells with a distinctive blue color when observed under a microscope, while viable cells appear unstained. Viable cells have intact cell membranes and hence do not take in dye from the surrounding medium. On the other hand, nonviable cells do not have an intact and functional membrane and hence do take up dye from their surroundings. This results in the ability to easily distinguish between viable and nonviable cells, since the former are unstained, small, and round, while the latter are stained and swollen. The method does not differentiate between apoptotic and necrotic cells.

The traditional method of performing trypan blue cell viability analysis involves manual staining and use of a hemocytometer for counting (1). Recent advances in instrumentation have led to a number of semi- or fully automated systems that can increase

Martin J. Stoddart (ed.), *Mammalian Cell Viability: Methods and Protocols*, Methods in Molecular Biology, vol. 740,
DOI 10.1007/978-1-61779-108-6_2, © Springer Science+Business Media, LLC 2011

the throughput and accuracy of this technique. The Beckman Coulter Vi-CELL™ XR Cell Viability Analyzer (2) will be described here, but there are numerous other instruments available, including the Nexcelom Bioscience Cellometer Vision® (3), the Invitrogen Countess™ Automated Cell Counter (4), and the Roche Innovatis Cedex (5). In the case of the Vi-CELL™, cell viability can be calculated and reported in percentage, concentration, and cell count, and additional parameters, such as cell diameter and circularity, are also provided. This instrument has been validated against manual cell counting and shown to be highly effective (6).

2. Materials

2.1. Manual Method

1. Trypan blue solution: 0.4% prepared in 0.81% sodium chloride and 0.06% dibasic potassium phosphate (see Note 1).
2. Bright-line hemocytometer and cover slip (Hausser Scientific, Horsham, PA) (see Note 2).
3. 70% (v/v) ethanol.
4. Microscope with 100× (10× eyepiece and 10× objective) magnification.
5. Micropipette and tips for 15 μL samples.

2.2. Automated Method

1. Vi-CELL™ XR Cell Viability Analyzer (Beckman Coulter, Inc., Fullerton, CA).
2. Vi-CELL™ XR Quad Pak Reagent Kit (Beckman Coulter, Inc.; see Note 3).
3. Vi-CELL™ Sample Vials (Beckman Coulter, Inc.).

3. Methods

The traditional cell counting method uses a hemocytometer (originally used for counting blood cells), which is a thick, glass microscope slide with an indented chamber of precise dimensions. This allows a defined volume of cell suspension to be deposited in the chamber, where the cells can be counted.

3.1. Manual Method of Sample Preparation

1. Prepare a suspension of approximately 1×10^6 cells/mL (see Note 4). Ensure that the sample is thoroughly mixed.
2. Make a 1:1 mixture of the cell suspension and the 0.4% trypan blue solution. The sample can be as small as 10 μL to several mL in volume. Gently mix and let stand for 5 min at room temperature.

3. Prior to use, wash the hemocytometer with 70% (v/v) ethanol and allow to dry.
4. Wash a coverslip with 70% (v/v) ethanol, allow to dry, and place on top of the hemocytometer counting chamber.
5. Apply 15 μL of cell suspension to the edge of the chamber between the cover slip and the V-shaped groove in the chamber. Allow the cell suspension to be drawn into the chamber by capillary action (see Note 5).
6. Let sit for 1–2 min and then count (see Note 6).

3.2. Manual Method of Cell Counting

The counting chamber of a hemocytometer is delineated by grid lines that identify the chamber areas to be used in cell counting. It has a depth of 0.1 mm and the four corner regions are typically used for cell counting. Each of these corner regions is 1 mm × 1 mm in dimension and is divided into 16 small squares in a four-by-four array (see Fig. 1). The volume of each of these four corners is thus 0.1 mm^3 or 1×10^{-4} mL.

1. Focus on the grid using the 10× objective.

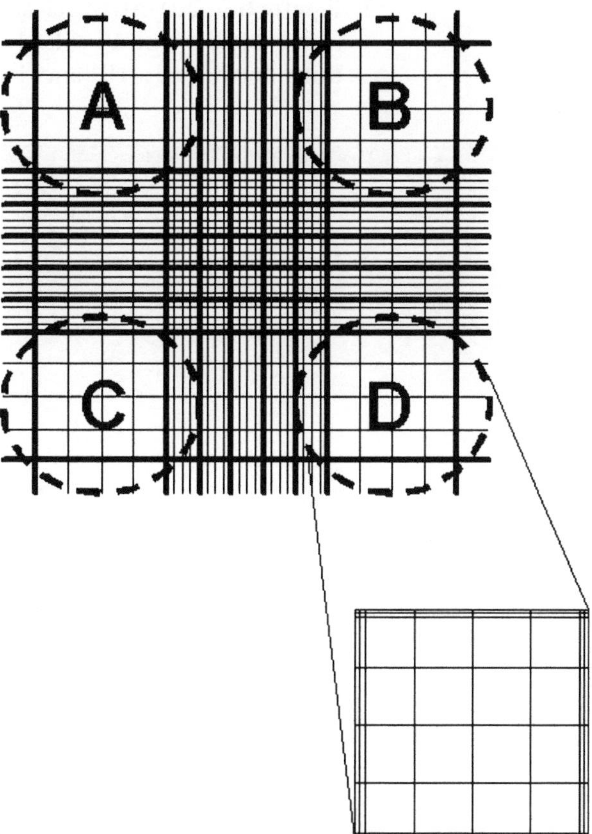

Fig. 1. The rulings on a hemocytometer.

2. Count the number of cells – both viable (unstained) and nonviable (stained) in each of the four corner quadrants (A, B, C, D) (see Note 7). Take the average of these four readings and multiply by 10^4 to obtain the number of cells per mL in the sample applied to the hemocytometer.
3. Multiply by two to take into account the 1:1 dilution of the sample in the trypan blue.
4. Multiply by any dilutions in the original sample preparation of the cell suspension.

$$\textit{Number of cells}\left(\textit{viable or}\ \text{nonviable}\right) = \frac{(A+B+C+D)}{4} \times 10^4 \times 2 \times \textit{sample dilution}$$

5. The percentage of unstained cells represents the percentage of viable cells in the suspension.

$$\%\ \text{Viable cells} = \frac{\text{Number of viable cells}}{\text{Total number of cells}}$$

3.3. Automated Method

1. Turn on the computer, instrument, and software program in this order.
2. Pipette 0.5–2.0 mL of cell suspension(s) into the sample vial(s) (see Note 8). The Vi-CELL™ requires sample cell concentrations between 5×10^4 and 1×10^7 cells/mL (see Note 9).

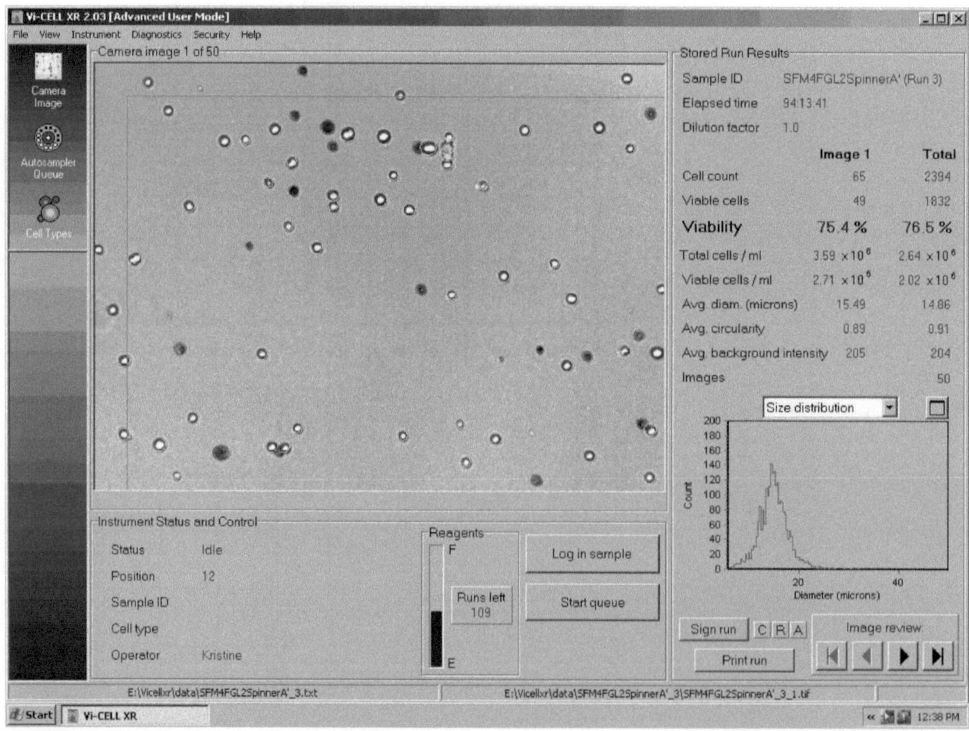

Fig. 2. The main window of the Vi-CELL XR 2.03 software program (Beckman Coulter) showing command icons, an image of the cell counting field, and a report of measured parameters.

3. Place the sample vials into vacant positions on the 12-position carousel.
4. Log in the sample on the software program.
5. Start the analysis by clicking on the "Start queue" icon (see Fig. 2).
6. Results of cell viability will be displayed on the computer screen (see Fig. 2).

4. Notes

1. Commercial preparations of trypan blue are usually filtered to remove debris. If the preparation is old or is made in the laboratory, it should be filtered through a 0.2-μm filter before use.
2. There are a number of commercial sources of hemocytometers. One of the most common ones in use is the improved Neubauer type used in this protocol.
3. The Vi-CELL™ XR Quad Reagent Pak kit consists of four solutions that are automatically dispensed during analysis: 0.4% trypan blue, buffer, cleaning agent, and 70% isopropyl alcohol disinfectant.
4. Adherent cells must first be treated with trypsin to create a cell suspension. A concentration of 1×10^6 cells/mL will result in an average of 50 cells per corner grid of the hemocytometer, which is a reasonable cell count for accuracy and precision. A concentration of 1×10^5 cells/mL will be too low to produce an accurate count. This sample should be concentrated by centrifugation before counting. A concentration of 1×10^7 cells/mL will be too high to count accurately and should be diluted before counting.
5. Some slight leakage out of the cover glass will ensure that the space under the cover glass is completely filled with cell suspension. Consistent application of the same (15 μL or other) volume will also help to ensure consistent results.
6. Do not leave cells in trypan blue for more than 15 min, in order to prevent cell death due to trypan blue toxicity.
7. If cells lie on the outside grid lines demarking the four corner quadrants, a consistent counting strategy should be adopted. It is common to include cells on the top and left lines and to exclude cells lying on the bottom and right lines.
8. The Vi-CELL™ has been used successfully to count mouse lymphocytes, Chinese Hamster ovary cells, human peripheral blood mononuclear cells, and mouse A9 fibroblast cells.

9. One of the advantages of the Vi-CELL™ is its ability to handle samples of varying concentration. A disadvantage is the requirement for a minimum of 0.5 mL of sample.

References

1. Cell and Tissue Culture: Laboratory Procedures. (1995) (Doyle, A., Griffiths, J.B., and Newell, D.G., Ed.). John Wiley & Sons, Inc., Chichester, England.
2. Beckman Coulter, Inc. (2004) V*i*-CELL™ XR Cell Viability Analyzer: Reference Manual. 1-82.
3. Nexcelom Bioscience. (2008) Cellometer® Vision: Automatic Cell Counter with Fluorescence Detection.
4. Invitrogen. (2008) Countess™ Automated Cell Counter.
5. Innovatis Inc. Cedex. Web site: http://www.innovatis.com/pdf/download/brochures/Cedex.pdf.
6. Louis, K.S., Siegel, A.C., and Levy, G.A. (2007) Comparison of Manual versus Automated Trypan Blue Dye Exclusion Method for Cell Counting. Web site: http://www.ibdl.ca/Application%20Notes/Application%20Note%20-%20Vi-CELL1.pdf.

Chapter 3

Estimation of Cell Number Based on Metabolic Activity: The MTT Reduction Assay

László Kupcsik

Abstract

The MTT reduction assay is used to determine the level of metabolic activity in eukaryotic cells, including animal, plant, and fungal cells. If the metabolic rate is constant, the technique can be employed to count living cells in a sample. Once it is set up, the method is very robust, and can be automatized to be applied on a large number of samples.

Key words: Cell count, Oxidoreductases, Intracellular space, MTT formazan, Tetrazolium salts, Energy metabolism

1. Introduction

Tetrazolium salts have been used to determine the reductive capacity of tissue homogenates since the 1940s (2, 3). The determination is based on the color change of tetrazolium salts as they undergo reduction. A type of tetrazolium, 3-(4,5-dimethylthiazo l-2-yl)-2,5-diphenyltetrazolium bromide (MTT), was used by Mosman to establish an automated cell counting system, in which the cells are assayed in the cell culture medium, without the need to wash or otherwise treat them (4). It has become a popular screening method for toxicity of pharmaceuticals and anticancer drugs. Interestingly, our understanding of the cellular mechanisms involved in the assay has been rather poor for decades. This was probably due to the extraordinary robustness of the method: very few researchers bother to study something that works. Later, however, it was necessary to thoroughly characterize the exact mechanism of MTT reduction partly because of its widespread use in standardized drug screening.

Martin J. Stoddart (ed.), *Mammalian Cell Viability: Methods and Protocols*, Methods in Molecular Biology, vol. 740, DOI 10.1007/978-1-61779-108-6_3,

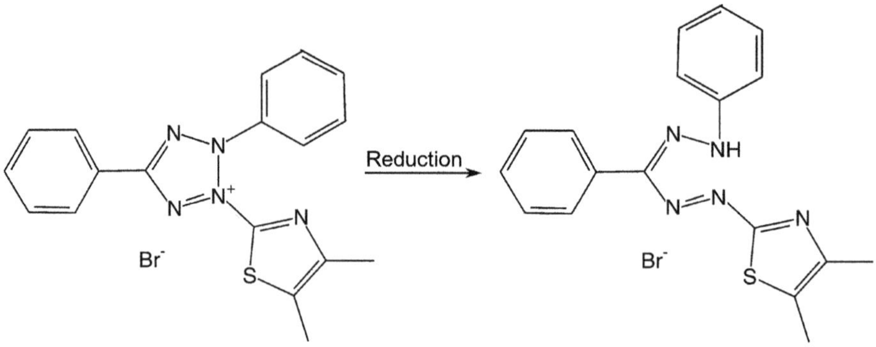

Fig. 1. Reduction of MTT.

1.1. Mechanism

The chemical basis of the assay is the reduction of the MTT (Fig. 1). MTT is a slightly yellow substance, which forms a blue formazan upon reduction.

The process primarily takes place in the cytoplasm, and to a lesser extent in the mitochondria and cell membrane (5–7). The reductase activity in the ER is highly dependent on the concentration of intracellular NADH and NADPH. The abundance of these nucleotide cofactors is associated with the availability of extracellular glucose. Hence, exhausted cell culture medium may cause lower MTT absorbance readings due to its low glucose concentration. The mitochondrial succinate dehydrogenase and cytochrome *c* take part in MTT reduction as well (8). Therefore, any substance or treatment that interferes with these enzymes or with glycolysis, changes the rate of MTT reduction, and consequently alters the result of cell counting.

As a consequence of these metabolic processes, dark purple needle-like formazan crystals appear, radiating from the cells in a few hours. Formazan crystals can be solubilized by mixing thoroughly in different organic solvents, mainly alcohols. First, ethanol was used for this purpose, but it caused precipitation of proteins in the culture medium. Isopropanol was found to be an equally good solvent and in most cases it lacks the adverse effects of ethanol (4). Detergents (10% sodium-dodecyl-sulfate [SDS], Triton X-100, Nonidet P-40) were also tested, and can be used when isopropanol is incompatible with the application (9). These latter compounds work without thorough mixing with the medium, but at the cost of an overnight incubation as opposed to few hours. Dimethyl sulfoxide was also reported to dissolve formazan effectively (10).

The absorption of the solubilized crystals can be measured around a broad peak wavelength between 570 and 590 nm. The use of a reference wavelength above 650 nm is recommended. The absorption is proportional to the cell number in an exceptionally wide linear range. As few as 200–1,000 cells are detectable in a 96-well plate, and the assay stays linear up to 50,000–100,000 cells per well (depending on cell type). The linear range of the assay should be tested by measuring serial dilutions of each new cell type used.

1.2. Alternatives

In certain cases, MTT may be replaced by other formazans, such as MTS, XTT, or WST-1, which are discussed in more detail elsewhere. They are generally more sensitive, less cytotoxic, and require fewer experimental steps (6, 7). On the other hand, they are affected by several factors (such as DTT, mercaptoethanol, l-cysteine, l-ascorbic acid, cyanide, azide, changes in environmental oxygen), to which MTT remains insensitive (6). Furthermore, MTT is metabolically reduced by *all* viable cells of every cell type, while others are not (1). MTT also shows a very good correlation (difference below 5%) with radioactive cell counting methods, which further justifies its credibility (9).

1.3. Limitations

The main limitations of the MTT method are summarized in Table 1.

Table 1
Interfering factors

Factor	Type of interference	Solution
Glucose	Lack of glucose lowers MTT production through decreased glycolysis	Do not use exhausted medium (change it few hours before assay)
Protein precipitation	Organic solvents may cause precipitation of serum proteins, which interferes with the reading	Use different solvent
Phenol red	Phenol red interferes with reading, its absorbance is pH-dependent	Use acidified solvent or medium without phenol red
Incomplete solubilization	Formazan crystals do not dissolve, and this lowers the sensitivity of the assay	Increase shaking intensity and/or extraction time
Confluency	(Over)confluent cells may lower their metabolic rate, resulting in underestimation of cell number	Use lower initial seeding density if possible
Metabolic rate	The metabolic rate of the cells is changed by the treatment	Use standard curve for each treatment, or find alternative cell counting method

2. Materials

1. Cell culture incubator.
2. ELISA plate reader.
3. Orbital shaker.
4. Cell culture medium with phenol red (e.g., DMEM with 10% FBS).
5. Cell culture medium without phenol red (optional, see Table 1: phenol red).
6. Isopropanol.
7. If phenol red-free medium is not available: acidified isopropanol (0.04 M HCl in isopropanol).
8. Cell culture treated polystyrene plates (compatible with the ELISA reader).
9. PBS pH 7.4.
10. 5 mg/ml MTT in PBS, filtered to remove occasional impurities.

3. Methods

The protocol below is optimized for standard 24-well tissue culture plates, the surface area of each well is 1.9 cm. Volumes are given on a per well basis. The reagent amount should be adjusted proportionally to wells of different sizes.

3.1. Cell Culture

The source of cells to be assayed can be immortalized cell line or primary cells extracted from a tissue. In general, cell lines should be used at low passage numbers. Do not use cells directly after thawing from cryopreservation, as these cells need a few days to recover from freezing. After such time under appropriate cell culture conditions, they can be used for MTT experiments. Both adherent and suspension cells can be used with this assay. Adherent COS7 cells will be used as an example in the following protocol description (for other cell types, see Note 1).

COS7 cells are cultured in high-glucose DMEM medium with 10% FBS. Medium should be changed 2–3 times a week, to ensure sufficient nutrient levels for the cells. Confluent cultures are split 1:3 or 1:5 regularly, using trypsin–EDTA to detach them. Cells are seeded out at about 10^6 cells per 80 cm^2. Healthy and exponentially expanding cells can be plated for the assay after trypsinization.

3.2. Cell Plating

1. Seed cells homogeneously in 24-well tissue culture plates at a density of 20,000 COS7 cells per well and let them attach overnight (see Note 2).
2. Change medium the next day, introduce factors or treatment conditions of interest. For example, add a wide range of concentrations of a chemical substance to the medium to test its toxicity.
3. Culture cells for 48 h in the presence of the substance in a humidified cell culture incubator at 37°C, 5% CO (see Note 3).

3.3. Carry Out the Assay

1. In the *phenol-red free* version of the assay, wash cells with 0.5 ml PBS, and add 0.5 ml phenol-red free medium to the cells. It should contain the tested factors, just like the phenol-red containing one used before. When *using phenol-red* in the assay medium, replace old medium with fresh one (optional, see Note 4).
2. Add MTT stock solution to the cells (50 μl).
3. Incubate in a humidified cell culture incubator at 37°C for 4 h (see Note 5).
4. Dissolve formazan salt by the addition of 0.5 ml isopropanol. If the assay medium contains phenol-red, 0.5 ml *acidified* isopropanol should be used instead. Cover with aluminum foil and shake on an orbital shaker at room temperature, 150 cycles/min, for 1 h.
5. Read absorbance at a wavelength between 570 and 590 nm on an ELISA plate reader. Optionally, a wavelength above 650 nm can be used as reference wavelength, to avoid inaccuracies caused by changed path length, evaporation, fingerprints, etc.).
6. Calculate results within the linear range of the assay, the absorbance value $A_{570-590}$ (or the absorbance ratio $\frac{A_{570-590}}{A_{650}}$) is proportional to the cell number, and can be used to determine relative cell numbers of samples by simply dividing the value of each sample with the value of the reference sample.

4. Notes

1. Various cells have different growth characteristics and growth requirements. The cell culture conditions should be suited to each cell type. For more information on ideal conditions, please consult the cell distributor or the available literature. Metabolic rates of various cells also differ, therefore the MTT assay is unfeasible to make cell number comparisons between cell types.

2. The appropriate cell density is determined by the cells' growth rate and the length of the culture period required by the application. Generally, confluency should be avoided at the end-point of the assay, due to reasons mentioned above (see Table 1). Toxicity testing usually requires at least 2–3 days' exposure to the tested substance.
3. Culture times may vary from a few hours to several weeks depending on the application. It must be taken into account at the cell plating that some cell types do not survive below a threshold cell density. During extended culture periods, the cells may reach confluency, which may alter the metabolic rate. This should be avoided if possible.
4. If cells are overgrown, or cultured for a long time without media change, they may use up the glucose in the medium, which will alter their MTT reduction ability. In extreme cases, the pH drops and the phenol-red in the medium turns yellow (which should be avoided). In all of these cases, it is recommended to replace the old medium with fresh one a few hours before adding the MTT to the cells.
5. The ideal length of incubation should be determined for each application. Too short times yield too little formazan product, and thus, inaccurate results. The MTT is reduced at a constant rate in the next phase of the assay. Ideally, the end point should be in this range. For most applications 4 h are adequate, but in certain conditions (e.g., lower temperature) the incubation time may be prolonged. If the cells are incubated too long, the reaction reaches a plateau due to crystal formation and cell death or possibly the exhaustion of MTT. The readings become unreliable in both cases.

References

1. E.H. Smail, B.N. Cronstein, T. Meshulam, A.L. Esposito, R.W. Ruggeri, and R.D. Diamond. (1992) In vitro, candida albicans releases the immune modulator adenosine and a second, high-molecular weight agent that blocks neutrophil killing. *J Immunol*, 148(11):3588–3595
2. Ernest Kun and L.G. Abood. (1949) Colorimetric estimation of succinic dehydrogenase by triphenyltetrazolium chloride.kuhn. *Science*, 109(2824):144–146
3. M.M. Nachlas, S.I. Margulies, and A.M. Seligman. (1960) Sites of electron transfer to tetrazolium salts in the succinoxidase system. *J Biol Chem*, 235:2739–2743
4. T. Mosmann. (1983) Rapid colorimetric assay for cellular growth and survival: application to proliferation and cytotoxicity assays. *J Immunol Methods*, 65(1-2):55–63
5. M.V. Berridge and A.S. Tan. (1993) Characterization of the cellular reduction of 3-(4,5-dimethylthiazol-2-yl)-2,5-diphenyltetrazolium bromide (mtt): subcellular localization, substrate dependence, and involvement of mitochondrial electron transport in mtt reduction. *Arch Biochem Biophys*, 303(2):474–482
6. Michael V. Berridge, An S. Tan, Kathy D. McCoy, and Rui Wang. (1996) The biochemical and cellular basis of cell proliferation assays that use tetrazolium salts. *Biochemica*, 4:14–19
7. Tytus Bernas and Jurek Dobrucki. (2002) Mitochondrial and nonmitochondrial reduction of mtt: interaction of mtt with tmre, jc-1, and nao mitochondrial fluorescent probes. *Cytometry*, 47(4):236–242

8. T.F. Slater, B.Sawyer, and U.Straeuli. (1963) Studies on succinate-tetrazolium reductase systems. iii. points of coupling of four different tetrazolium salts. *Biochim Biophys Acta*, 77:383–393
9. H.Tada, O.Shiho, K. Kuroshima, M. Koyama, and K. Tsukamoto. (1986) An improved colorimetric assay for interleukin 2. *J Immunol Methods*, 93(2):157–165
10. S.A. Jabbar, P.R. Twentyman, and J.V. Watson. (1989) The mtt assay underestimates the growth inhibitory effects of interferons. *Br J Cancer*, 60(4):523–528

Chapter 4

WST-8 Analysis of Cell Viability During Osteogenesis of Human Mesenchymal Stem Cells

Martin J. Stoddart

Abstract

WST-8 is one of the newer generation formazan-based dyes, which release the converted product into the medium in a soluble form. This allows for a non-destructive determination of viability enabling the cells to be subject to further investigations. This is a major advantage in cases where cell phenotype is being investigated and data such as matrix synthesis is correlated to cell number. This chapter describes the use of WST-8 to normalise alkaline phosphatase activity.

Key words: Stem cells, Formazan, Matrix synthesis, Osteogenesis

1. Introduction

Formazan-based dyes have been routinely used to assess cell viability for many years. The most commonly used was MTT which was the mainstay of viability methods due to its ease of use and standardised protocols (1). As MTT conversion results in an insoluble precipitate which resides within the cells, the method requires the cells to be broken open and the precipitate to be dissolved in order to quantify to viability. This not only increases the number of steps required but as it is destructive it requires dedicated wells at each time point. The next generation of dyes were those that resulted in the release of a water-soluble formazan dye upon reduction, such as XTT, MTS, WST-1, and WST-8.

WST-8 is cleaved to formazan by cellular dehydrogenases. The amount of dye converted is directly proportional to the number of living cells and has been shown to correlate well with the ^{3}H-thymidine incorporation assay commonly used for proliferation studies (2).

Martin J. Stoddart (ed.), *Mammalian Cell Viability: Methods and Protocols*, Methods in Molecular Biology, vol. 740,
DOI 10.1007/978-1-61779-108-6_4, © Springer Science+Business Media, LLC 2011

WST-8 offers the advantage that it is stable and so is supplied as a single, ready to use reagent. It also has no harvesting, washing, or solubilisation steps. Finally, it has been reported to be more sensitive than MTT, XTT, MTS, or WST-1 (2). The absorbance is measured at 450 nm (with 650 nm as the reference) and as the toxicity of the reagent is low multiple measurements can be carried out on the same plate over time.

In cases where cell phenotype is being investigated, data such as matrix synthesis are correlated to cell number. As WST-8 is a non-destructive method, the same samples can be used for further measurements, thus reducing the number of initial samples required. In the case of osteogenic induction of stem cells, the level of alkaline phosphatase (ALP) activity is commonly used as an indicator of phenotype. The assay is based on the hydrolysis of *p*-nitrophenyl phosphate in an alkaline buffer by ALP, yielding *p*-nitrophenol and inorganic phosphate (3). Addition of NaOH stops the reaction and converts *p*-nitrophenol to a yellow complex readily measured at 400–420 nm. The intensity of colour formed is proportional to phosphatase activity over a linear range.

2. Materials

2.1. Cell Culture

1. Human bone marrow-derived mesenchymal stromal cells.
2. Complete αMEM: αMEM containing 10% FCS and 1× penicillin/streptomycin.
3. Osteogenic medium: Complete αMEM containing 1:100 non-essential amino acids (from 100× stock), 10^{-7} dexamethasone, 10 mM β-glycerophosphate, and 50 μg/ml ascorbic acid.
4. WST-8 reagent.
5. 24-Well cell culture treated plates.
6. 1× Sterile phosphate-buffered saline (PBS).

2.2. Quantitative Alkaline Phosphatase Activity Determination

1. 0.1% Triton-X in 10 mM Trizma base, adjusted to pH 7.4 with HCl.
2. Phosphatase substrate: 4-Nitrophenyl phosphate disodium salt hexahydrate (Sigma N2765, 20 mg tablets, stored at –20°C).
3. Standard solution: *p*-Nitrophenol 10 mM (store at 4°C).
4. Alkaline buffer stock solution: 10.48 M 2-Amino-2-methylpropanol (store at 4°C).
5. Protease inhibitor cocktail (Sigma, P8340).
6. Diethanolamine working buffer: Prepare 1 M diethanolamine in 0.5 mM $MgCl_2$: add 424 μl of diethanolamine to 4 ml of

0.553 mM $MgCl_2$ (112.5 mg/l), pH to 9.8. Final concentration will be 1 M diethanolamine and 0.5 mM $MgCl_2$.

7. Alkaline buffer: Dilute the stock (2-amino-2-methylpropanol) to a final concentration of 1.5 M with H_2O, and adjust to pH 10.3 at 25°C. Store at 4–8°C.
8. Substrate solution: Dissolve five 4-nitrophenyl phosphate disodium salt hexahydrate tablets (20 mg each) in 3.8 ml of the diethanolamine working buffer to a final concentration of 25 mg/ml. The substrate solution can be stored at –20°C for 6 weeks in a light-shielded tube.
9. Plate reader with filter at 405, 450, and 650 nm.

3. Methods

3.1. Cell Culture

1. Plate human bone marrow-derived stem cells at 16,000 cells/cm^2 in 24-well plates in 1 ml complete αMEM (see Note 1). Allow to adhere overnight at 37°C and 5% CO_2.
2. Following day, remove medium and replace with 1 ml osteogenic medium. Culture at 37°C and 5% CO_2.
3. Re-feed cells every 2 days with fresh osteogenic medium.
4. After 14 days of culture measure cell viability using WST-8.

3.2. WST-8 Assay

1. Add 100 µl of WST-8 reagent to each well and incubate for 1–4 h (see Note 2). Alternatively, prepare fresh osteogenic medium containing 10% WST-8 reagent, pre-warm to 37°C, and add 1 ml to cells, followed by a 1–4-h incubation.
2. Measure the absorbance at 450 nm in a microplate reader. This measurement can be repeated later if required.

3.3. Quantitative Alkaline Phosphatase Activity Determination

1. Prior to carrying out ALP activity wash each well 3× 5 min with PBS.
2. To each well add 300 µl of 0.1% Triton-X in 10 mM Tris–HCl, pH 7.4 and protease inhibitor cocktail (1:1,000). Store at 4°C under agitation for 1–3 h (depending on the activity of the sample).
3. Harvest sample into a 1.5-ml Eppendorf tube.
4. Samples are centrifuged 1,000 × *g* for 10 min and supernatant is transferred into a fresh tube, the pellet can be discarded (see Note 3).
5. Prepare a standard curve by diluting the *p*-nitrophenol standard (10 mM) 1:10 with 0.1 % Triton-X in 10 mM Trizma base to a final concentration of 1 mM (see Note 4). From the diluted standard solution 0, 10, 20, 30, 40, 50, 60, and 70 µl is added to Eppendorf tubes and made up to 100 µl with 0.1 %

Triton-X in 10 mM Trizma base. The standards are then treated as the samples.

6. Cell lysate samples are thawed on ice and 200 µl (if high values expected, only 100 µl used + 100 µl deionised water) per sample and only 100 µl standards (+100 µl deionised water) are added to Eppendorf tubes.
7. Add 250 µl of alkaline buffer and 50 µl substrate. During this addition the reaction mix is kept on ice.
8. Start the reaction simultaneously for all samples by incubating at 37°C. (see Note 5).
9. Incubate the samples and standards for exactly 15 min. Weak samples can be incubated up to 45 min and activity expressed as activity per minute (see Note 6).
10. Inactivate the enzyme by the addition of 500 µl 0.1 M NaOH.
11. Measure each sample and standard in duplicate (250 µl each) at 405 nm absorbance using a 96-well plate reader.
12. Add 50 µl of concentrated HCl to each well, and measure again. Subtract the second measurement from the first, to remove interference.
13. Divide the measure ALP activity by the measured WST-8 absorbance to determine ALP activity as a factor of cellular activity.

4. Notes

1. Various initial cell seeding densities have been reported for osteogenic induction protocols. Due to the nature of the WST-8 it is critical that the amount of reagent added is sufficient for the number of viable cells within the well. Preliminary experiments should be carried out to confirm that the samples are in the linear range of the assay.
2. Too long incubation times can lead to the depletion of the reagent and the assay would no longer be in the linear range. Ensure that the WST-8 reagent is in excess at all times.
3. Samples can be stored frozen at –20°C until analysis.
4. Due to the light sensitivity of the NBT substrate all further steps are performed by avoiding direct light exposure.
5. During the incubation the reactions will develop a yellow colour which correlates with the activity of the sample ALP.
6. All samples that are compared to each other should be incubated for the same amount of time for consistency.

References

1. T. Mosmann. (1983) Rapid colorimetric assay for cellular growth and survival: application to proliferation and cytotoxicity assays. *J Immunol Methods*, **65**(1-2):55–63
2. Tominaga H., Ishiyama M., Ohseto F., Sasamoto K., Hamamoto T., Suzuki K., and Watanabe M. (1999) A water-soluble tetrazolium salt useful for colorimetric cell viability assay. *Anal. Commun.* **36**: 47–50
3. Farley J.R., Hall S.L., Ilacas D., Orcutt C., Miller B.E., Hill C.S., Baylink D.J., (1994) Quantification of skeletal alkaline phosphatase in osteoporotic serum by wheat germ agglutinin precipitation, heat inactivation, and a two-site immunoradiometric assay. *Clin.Chem.* **40**: 1749–1756.

Chapter 5

Assessment of Cell Proliferation with Resazurin-Based Fluorescent Dye

Ewa M. Czekanska

Abstract

The Alamar Blue assay is based on enzymatic reduction of indicator dye by viable cells and serves as an effective tool for assessing cell proliferation and as a screening technique. It can be applied in studies concentrating on animal, plant, yeast, and bacteria cells. Among the various methods for cell viability and cytotoxicity, it utilises all features of ideal and reliable test; it is one-step, sensitive, safe, non-toxic for cells, and cost-effective.

Key words: Cell proliferation, Resazurin, Alamar Blue, Metabolic activity, Oxidoreductases

1. Introduction

The resazurin-based assay was first introduced in the late 1920s to investigate the sanitary condition of milk (1–3). Later, it was applied to plant metabolism studies (4), for assessing semen quality (5) and antifungal susceptibility testing (6). Due to many advantages of resazurin-based assay, it has also become a useful tool for investigations of toxicants (7–9). Moreover, the simplicity, safety, homogenous nature, and sensitivity give this assay a predominant position over the other classic tests used for estimating cell viability and proliferation (10, 11).

There are many assays available in the market that are composed of resazurin sodium salt dye to monitor in vitro mammalian cell proliferation, such as Alamar Blue (Alamar Biosystems; Invitrogen; AbD Serotec; Arcus), CellTiter Blue (Promega) and Cell toxicity Colorimetric/Fluorometric Assay (Sigma–Aldrich). The resazurin is a blue weakly fluorescent indicator dye that changes into highly fluorescent pink resorufin in response to irreversible chemical reduction (Fig. 1).

Martin J. Stoddart (ed.), *Mammalian Cell Viability: Methods and Protocols*, Methods in Molecular Biology, vol. 740,
DOI 10.1007/978-1-61779-108-6_5,

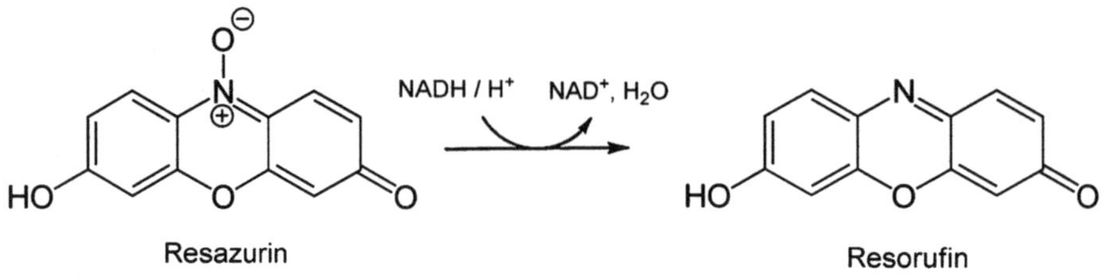

Resazurin

NADH / H+ NAD+, H_2O

Resorufin

Fig. 1. Mechanism of reduction of resazurin to resorufin.

Inside the cell, Alamar Blue undergoes enzymatic reduction in mitochondria due to the activity of enzymes such as: flavin mononucleotide dehydrogenase, flavin adenine dinucleotide dehydrogenase, nicotinamide adenine dehydrogenase, and cytochromes (12). It was also noted that cytosolic and microsomal enzymes have abilities to reduce resazurin. Moreover, the extent of mitochondrial reduction is similar to cytosolic reduction driven by NADPH:quinine oxidoreductase, flavin reductase, and cytochromes (13). The red resorufin is excreted outside the cells to the medium which results in visible colour change from blue to pink. Hence, the rate of reduction based on colour change, which can be quantified colorimetrically or fluorometrically, reflects the number of viable cells. Alamar Blue is very sensitive and, depending on cell type and incubation time, is linear in the range of 50–50,000 cells in 96-well plate.

The maximum absorbance for resazurin/resorufin is 605/573 nm, whereas the maximum peak of excitation/emission spectra for resorufin is 579/584 nm. It is recommended to carry out the measurements using fluorescence as it requires fewer calculations and is more sensitive due to considerable overlap of the absorbance spectra for oxidised and reduced forms of dye. To monitor the reduction process of Alamar Blue dye by cells wide range of fluorescence filters can be used: 530–570 nm for excitation and 580–620 nm for emission.

2. Materials

1. Cell culture incubator.
2. Fluorescence plate reader with excitation 530–570 nm and emission 580–620 nm filter pair.
3. Osteoblastic cell line MC3T3-E1 cells.
4. Cell culture medium (DMEM with 10% FBS and 1% antibiotics).
5. Cell culture treated polystyrene 24-well plates.

6. 96-Well opaque-walled white plate.
7. 0.1 M phosphate-buffered saline (PBS) solution, pH 7.4.
8. Alamar Blue.
9. Light-shield tubes or 15-ml centrifuge tubes wrapped in aluminium foil.

3. Methods

The following protocol is optimised for 24-well tissue culture plates with 1.9 cm^2 surface area. Volumes of reagents are given per well and they should be adjusted accordingly if other sizes are used. The Alamar Blue assay can be optimised and applied to various cell types (see Note 1), including primary cells obtained after extraction from tissue and immortalised cell line. For the assay, exponentially expanded cells should be plated after trypsinisation. It is not recommended to use cells directly after thawing from cryopreservation. As an example in this protocol MC3T3-E1 cells cultured in DMEM with 10% FCS and 1% antibiotics will be used (see Note 2).

3.1. Cell Plating for Standard Curve

1. Seed cells in 24-well tissue culture plates with 500 μl of medium at the following densities: 1,000; 2,000; 5,000; 10,000; 20,000; 40,000; 80,000; 100,000. Include four samples for each density.
2. Transfer plates to the incubator (37°C with an atmosphere of 5% CO_2 and 95% humidity) and let them attach for 5 h. After that time check if all cells are attached to the surface of the well; if not incubate longer till all cells are attached.
3. Prepare 10% solution of Alamar Blue in culture medium in a light-shield tube (see Notes 3 and 4). Calculate the amount needed by multiplying the amount of samples by the volume of solution per well (300 μl). In calculations include three no-cell controls.
4. Carefully aspirate medium from each well and add 300 μl of fresh medium containing a 10% solution of Alamar Blue to the wells.
5. Wrap culture plates in aluminium foil and incubate for 4 h in cell culture incubator at 37°C with an atmosphere of 5% CO_2 and 95% humidity.
6. Transfer a 150-μl aliquot of each sample and media with dye alone to a 96-well plate in duplicates.
7. Read fluorescence at excitation 530–570 nm and emission 580–620 nm wavelength on plate reader (see Note 5).

8. Calculate the results by averaging the values obtained for each density and subtracting the average value of no-cell control. Plot the results of fluorescence/cell number and define the equation. For MC3T3-E1 cells the following equation was obtained: $f(x) = 0.0587x + 138.14$.

3.2. Cell Plating for Estimating Proliferation

1. Seed 10,000 cells per well in 24-well tissue culture plates with 500 μl of medium (four samples for each time point; see Note 6).
2. At each time point, aspirate the media from four wells and wash samples with 500 μl of 0.1 M PBS solution (see Note 7).
3. Add 300 μl of fresh medium containing a 10% solution of Alamar Blue to the wells (see Note 8). Include three no-cell controls.
4. Proceed according to steps 5–7 in previous section.
5. Calculate the results by averaging fluorescent values of the samples and subtracting average value of samples without cells.
6. Calculate cell number at each time point using the equation defined by standard curve.

4. Notes

1. Before performing the assay for the first time, optimise experimental parameters, such as the incubation time, amount of cells, and amount of Alamar Blue used. This is very important as various cell types have unique metabolic capacity to reduce resazurin to resorufin. Here, predetermined 4-h incubation time and 10% solution of the dye were used for densities 1,000–100,000 cells/well. However, these conditions may not be optimal for other cell types. Hence, perform the screening assays with various cell densities, at least three concentrations of Alamar Blue and incubation times. Incubating cells for too long with resazurin solution results in a secondary reduction of pink fluorescent resorufin into non-fluorescent colourless hydroresorufin which further leads to aberrant and inaccurate results. Table 1 illustrates the empirically determined requirements for various cell lines (14).
2. Resazurin can be reduced by commonly known antioxidants, such as ascorbic acid, cysteine, dithionite, and dithiothreitol (15). When using culture medium supplemented with ascorbic acid the optimal incubation time may vary from culture in the same type of medium without the ascorbic supplementation.
3. Depending on manufacturer Alamar Blue can be stored at −20°C, 4°C, or at room temperature. Generally, lower temperatures increase the product stability. If −20°C storage conditions

Table 1
The recommended conditions for the Alamar Blue assay in 96-well plate for normal and cancer cell lines. Reprinted from (14) with permission from Elsevier

Cell line	Linear range (cell number × 10^4)	Optimal AB concentration (%)	Optimal incubation time (h)
Normal cell lines			
BALB/3T3	0.05–2	4	3
CHO-K1	0.05–3	10	3
NJTIDF 4054	0.05–2	4	3
Ovarian cancer cell lines			
2008	0.05–3	4	1
IGROV-1	0.05–3	10	1
OVCAR-3	0.05–3	4	3
SK-OV-3	0.05–3	4	3
Leukaemia cell lines			
HL-60	0.5–5	10	6
K-562	0.5–5	10	6
MOLT-4	0.5–5	10	6

are used, reduce the freeze–thaw cycles by 10. For this, the best is to aliquot the stock solution.

4. Resazurin and resorufin are light-sensitive and need to be protected from light otherwise it results in decreased sensitivity.
5. It is recommended to read the plate on the day an experiment is performed. If it is not possible, 96-well plate with sample aliquots can be stored at 4°C wrapped in foil and measurement can be taken within 1–3 days without affecting fluorescence or absorbance. However, when measuring fluorescence, it has to be kept in mind that temperature affects the fluorescence values. Thus, to keep the conditions of fluorescence readings constant wait till refrigerated plate is warmed up to the ambient temperature before reading.
6. Optionally, the reaction can be stopped and stabilised after the incubation by adding 50 μl of 3% SDS in PBS (pH 7.4) per 100 μl of original culture volume. Plates can be stored at ambient temperature wrapped in foil up to 24 h.
7. Although Alamar Blue is known as a non-toxic in vitro assay in short-time exposure, the prolonged contact of cells with resazurin may affect the reduction rate and cell viability (16). Thus, it is possible to perform more than one type of assay

on the same sample, but for proliferation assessment it is better to use Alamar Blue as end point assay.

8. Alamar Blue should not be added directly to cultured cells in long-time experiments as protein concentration may affect the results by quenching the fluorescence signal (17). The Alamar Blue solution at determined concentration should be prepared freshly each time and before applying, samples should be washed with PBS to reduce the risk of different protein concentrations generated during the culturing time.

References

1. Twigg R.S. (1945) Oxidation-reduction aspects of resazurin. *Nature,* **155**:401–402
2. Nixon M. C. and Lamb A. B. (1945) Resazurin Test for Grading Raw Milk. *Can J Comp Med Vet Sci,* **9**(1):18–23
3. Ali-Vehmas T., Louhi M., and Sandholm M. (1991) Automation of the resazurin reduction test using fluorometry of microtitration trays. *Zentralbl Veterinarmed B,* **38**(5): 358–372
4. De Jong D. W. and. Woodlief W. G. (1977) Fluorimetric assay of tobacco leaf dehydrogenases with resazurin. *Biochim Biophys Acta,* **484**(2):249–259
5. Glass R. H., Ericsson S. A., Ericsson R. J., Drouin M. T., Marcoux L. J., and Sullivan H. (1991) The resazurin reduction test provides an assessment of sperm activity. *Fertil Steril,* **56**(4):743–746
6. To W. K., Fothergill A. W., and Rinaldi M. G. (1995) Comparative evaluation of macrodilution and alamar colorimetric microdilution broth methods for antifungal susceptibility testing of yeast isolates. *J Clin Microbiol,* **33**(10):2660–2664
7. O'Brien J., Wilson I., Orton T., and Pognan F. (2000) Investigation of the Alamar Blue (resazurin) fluorescent dye for the assessment of mammalian cell cytotoxicity. *Eur J Biochem,* **267**(17):5421–5426
8. Perrot S., Dutertre-Catella H., Martin C., Rat P., and Warnet J. M. (2003) Resazurin metabolism assay is a new sensitive alternative test in isolated pig cornea. *Toxicol Sci,* **72**(1):122–129
9. Hamid R., Rotshteyn Y., Rabadi L., Parikh R., and Bullock P. (2004) Comparison of alamar blue and MTT assays for high throughput screening. *Toxicol In Vitro,* **18**(5): 703–710
10. Larson E. M., Doughman D. J., Gregerson D. S., and Obritsch W. F. (1997) A new, simple, nonradioactive, nontoxic in vitro assay to monitor corneal endothelial cell viability. *Invest Ophthalmol Vis Sci,* **38**(10):1929–1933
11. Jonsson K. B., Frost A., Larsson R., Ljunghall S., and Ljunggren O. (1997) A new fluorometric assay for determination of osteoblastic proliferation: effects of glucocorticoids and insulin-like growth factor-I. *Calcif Tissue Int,* **60**(1):30–36
12. O'Brien, J., Wilson, I., Orton, T., and Pognan, F. (2000) Pognan. Investigation of the Alamar blue (resazurin) fluorescent dye for the assessment of mammalian cell cytotoxicity. *Eur. J. Biochem,* **267**, 5421–5426.
13. Gonzalez R. J. and Tarloff J. B. (2001) Evaluation of hepatic subcellular fractions for Alamar blue and MTT reductase activity. *Toxicol In Vitro,* **15**(3):257–259
14. Nakayama G. R., Caton M. C., Nova M. P., and Parandoosh Z. (1997) Assessment of the Alamar Blue assay for cellular growth and viability in vitro. *J Immunol Methods,* **204**(2):205–208
15. De Jong, D. W. and Woodlief, W. G. (1977) Fluorometric assay of tobacco leaf dehydrogenases with resazurin. *Biochim. Biophys. Acta,* **484**, 249–259.
16. Squatrito R. C., Connor J. P., and Buller R. E. (1995) Comparison of a novel redox dye cell growth assay to the ATP bioluminescence assay. *Gynecol Oncol,* **58**(1):101–105
17. Jonsson, K. B., Frost, A., Larsson, R., Ljunghall, S., and Ljunggren, O. (1997) A new fluorometric assay for determination of osteoblastic proliferation: effects of glucocorticoids and insulin-like growth factor-I. *Calcif. Tissue Int,* **60**, 30–36.

Chapter 6

The xCELLigence System for Real-Time and Label-Free Monitoring of Cell Viability

Ning Ke, Xiaobo Wang, Xiao Xu, and Yama A. Abassi

Abstract

We describe here the use of the xCELLigence system for label-free and real-time monitoring of cell viability. The xCELLigence system uses specially designed microtiter plates containing interdigitated gold microelectrodes to noninvasively monitor the viability of cultured cells using electrical impedance as the readout. The continuous monitoring of cell viability by the xCELLigence system makes it possible to distinguish between different perturbations of cell viability, such as senescence, cell toxicity (cell death), and reduced proliferation (cell cycle arrest). In addition, the time resolution of the xCELLigence system allows for the determination of optimal time points to perform standard cell viability assays as well as other end-point assays to understand the mode of action. We have used the WST-1 assay (end-point viability readout), the cell index determination (continuous monitoring of viability by xCELLigence), and the DNA fragmentation assay (end-point apoptosis assay) to systematically examine cytotoxic effects triggered by two cytotoxic compounds with different cell-killing kinetics. Good correlation was observed for viability readouts between WST-1 and cell index. The significance of time resolution by xCELLigence readout is exemplified by its ability to pinpoint the optimal time points for conducting end point viability and apoptosis assays.

Key words: Impedance, Cell viability, Cell index, Apoptosis, MG-132, 5-FU

1. Introduction

Monitoring of cell viability is critical to many areas of basic and biomedical research both from a mechanistic perspective to understand the molecular and biochemical pathways regulating cell viability and from a therapeutic angle to develop agents which modulate cell viability. Because cell viability is critical for overall function and physiology of the cell, tissue, and organ, it is subject to regulation by variety of cellular pathways mediated by extrinsic and intrinsic factors. Viability assays are routinely used to study

Martin J. Stoddart (ed.), *Mammalian Cell Viability: Methods and Protocols*, Methods in Molecular Biology, vol. 740, DOI 10.1007/978-1-61779-108-6_6,

how perturbation of these pathways elicits their effect on overall cell viability.

Traditional viability assays include the Trypan Blue dye exclusion assay to test cell membrane integrity and colony formation assays to test continuous growth of cells (1). The exclusion of Trypan Blue or propidium iodide dyes is often used as an indication of membrane integrity of living cells, as the dyes can cross the compromised cell membrane and stain cellular targets or structures in dead cells. The colony formation assay can document any effects on cell proliferation, even longer-term cell cycle arrest, or antiproliferative effects. More recently, homogeneous assays for cell viability have been developed which are amenable to automation for high-throughput screening (HTS). Viability markers, usually a biochemical event that occurs only in living cells, are often used for HTS assays. Tetrazolium salts, such as MTT, XTT, and WST-1, are converted to a formazan product by active mitochondria, which can then be measured as a colorimetric readout (2–4). Similarly, nonfluorescent resazurin, e.g., alamarBlue, can be converted to fluorescent resorufin by active mitochondria, which can then be measured by fluorescence (5–7). Cellular ATP content is also used as a cellular marker for viability (8–11). These assays are conducted at the end of the study to determine the difference in cellular viability between control and toxin-treated samples.

Modulation of cell viability can occur by a multitude of mechanisms and pathways. Cell death can occur through a spectrum of distinct morphological and biochemical pathways culminating in apoptosis, necrosis, or autophagy (12). Apoptosis is characterized by shrinking of the cytoplasm, nuclear condensation, and formation of apoptotic bodies, while necrosis involves cell swelling, break-down of the plasma membrane, and ultimately cell lysis. Antiproliferative drugs exert their functions by inhibiting cell division and reducing cellular proliferation. It is important to understand the mechanism(s) of cell death or reduced proliferation induced by drug treatment, in order to understand the mechanism of drug action and predict its possible side effects. However, the effect of a compound on viability is often transient and therefore very difficult to detect if not assayed at the optimal time points. For example, apoptosis occurs only during a short period of time, often within hours, and is frequently followed by secondary necrosis events. In some cases, induced cell cycle arrest may be temporary, while in other cases a prolonged arrest may occur which often results in cell demise. Considering that different toxins exert their maximal effects at different time points after treatment, it is useful to monitor viability continuously to pinpoint the optimal time points at which endpoint assays should be employed to decipher the mechanism of action of cytotoxicity or reduced proliferation.

Additional complicating factors include toxin concentration and cell lines tested, as the kinetics of cytotoxicity is often different between treatment conditions. For instance, higher toxin concentrations may exert cytotoxic effects with a faster kinetics, while lower concentration may have a slower antiproliferative effect or even promote proliferation, as in the case with digitalis (13). Furthermore, depending on the genetic backgrounds of the tested cell lines, same toxin may have very different effects, as in the cases with many target-directed therapeutics that showed much more robust effects in the so-called oncogene-addicted cell lines (14). Therefore, for each toxin, each individual concentration, and each cell line tested, it is important to optimize the assay time point(s) required to obtain the most meaningful results, which can be very labor-intensive.

The xCELLigence system allows for label-free and dynamic monitoring of cellular phenotypic changes in real time using impedance as readout. The system measures electrical impedance across interdigitated micro-electrodes integrated on the bottom of tissue culture E-Plates (Fig. 1a). The impedance measurement, which is displayed as cell index (CI) value, provides quantitative information about the biological status of the cells, including cell number, viability, and morphology. Impedance-based monitoring of cell proliferation/viability correlates very well with cell number and WST-1 readout (Figs. 1 and 2). In addition, continuous monitoring of cellular responses to biologically active small molecule compounds produces time-dependent cellular response profiles (TCRPs), that can be predictive of mechanism of action (15). For example, HeLa cells that were treated with a DNA-damaging agent (5-FU) and a proteasome inhibitor (MG-132) produced very specific TCRPs that are modulated as a function of time (Fig. 2). The kinetic cell viability measurement provides the temporal information as to when a drug of interest induces its cytotoxic effect, and can thus allow one to pinpoint the optimal time points (e.g., when the cytotoxicity reaches maximal effect, lowest cell index) for more specific cell fate assays, the effect of which can often be transient and difficult to capture. In the specific examples we have shown in this work, it is clear that 5-FU and MG-132 induce cytotoxicity with different kinetics, even though the ultimate fate of cells in response to both treatments is apoptosis (Figs. 2 and 3). For MG-132, apoptosis is observed at 16 h post compound addition (after cell index reached lowest level), but not at 64 h, confirming the transient nature of apoptotic process (Figs. 2a and 3a). In contrast, apoptotic induction by 5-FU is only observed at 64 h post treatment (after cell index reached lowest level), but not at 16 h (no cell index changes observed) (Figs. 2e and 3b). This example clearly demonstrates the importance of considering the time factor when performing viability or apoptosis assays.

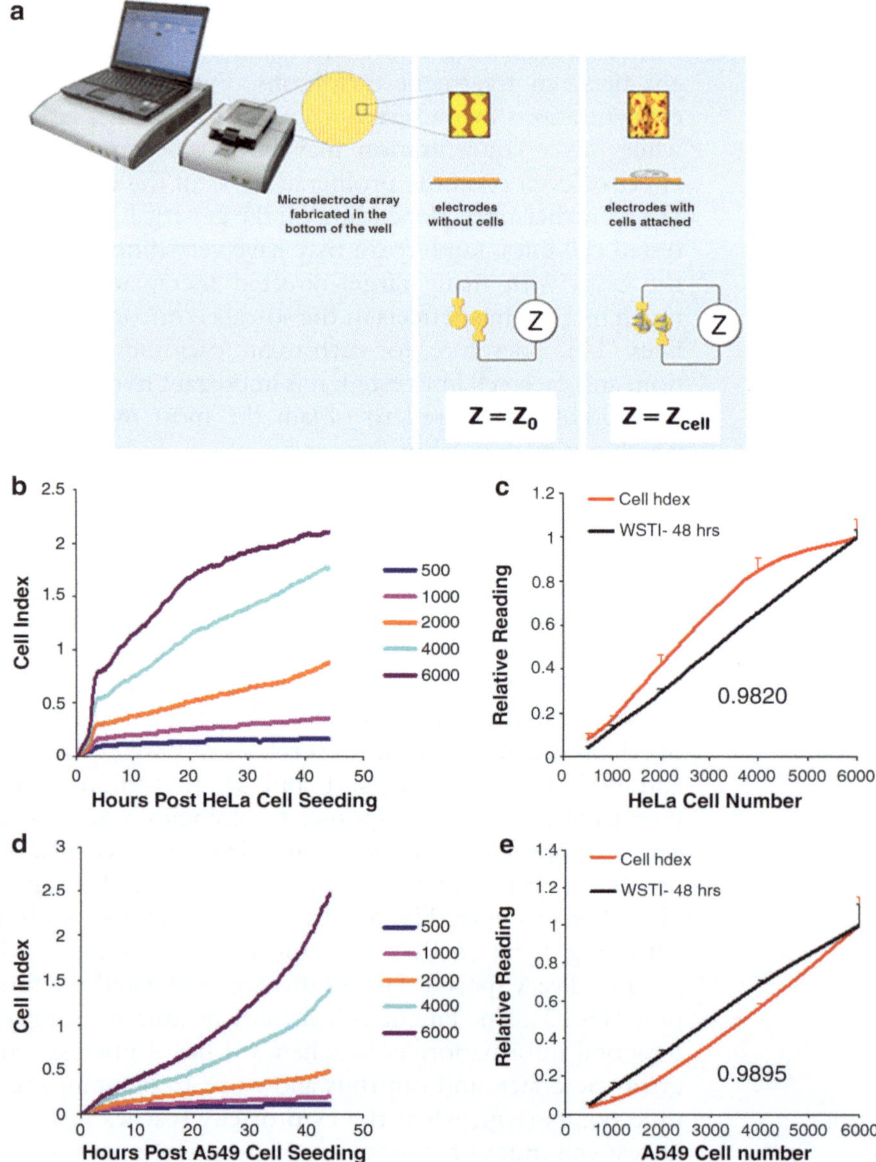

Fig. 1. xCELLigence in real-time monitoring of cell proliferation. (**a**) RTCA-SP instrument is shown, consisting of control unit, station, E-Plate, and display unit. Interdigitated gold electrodes at the bottom of the E-Plate are shown with or without cells. Background impedance is determined for medium alone (Z_0), then after the cells attach and proliferate on the electrodes (Z_{cell}). (**b–e**) Different numbers of HeLa (**b**) and A549 (**d**) cells were seeded in the E-Plate and the cell index was continuously monitored for 45 h. At the end of study, WST-1 was added to the medium and absorbance determined (**c** and **e**). Both cell index and WST-1 readings at the end of study were normalized against the value obtained at the highest seeding density. The average of triplicate samples is plotted with error bars indicating standard deviation. The Pearson correlation coefficient between WST-1 and cell index readings is also indicated.

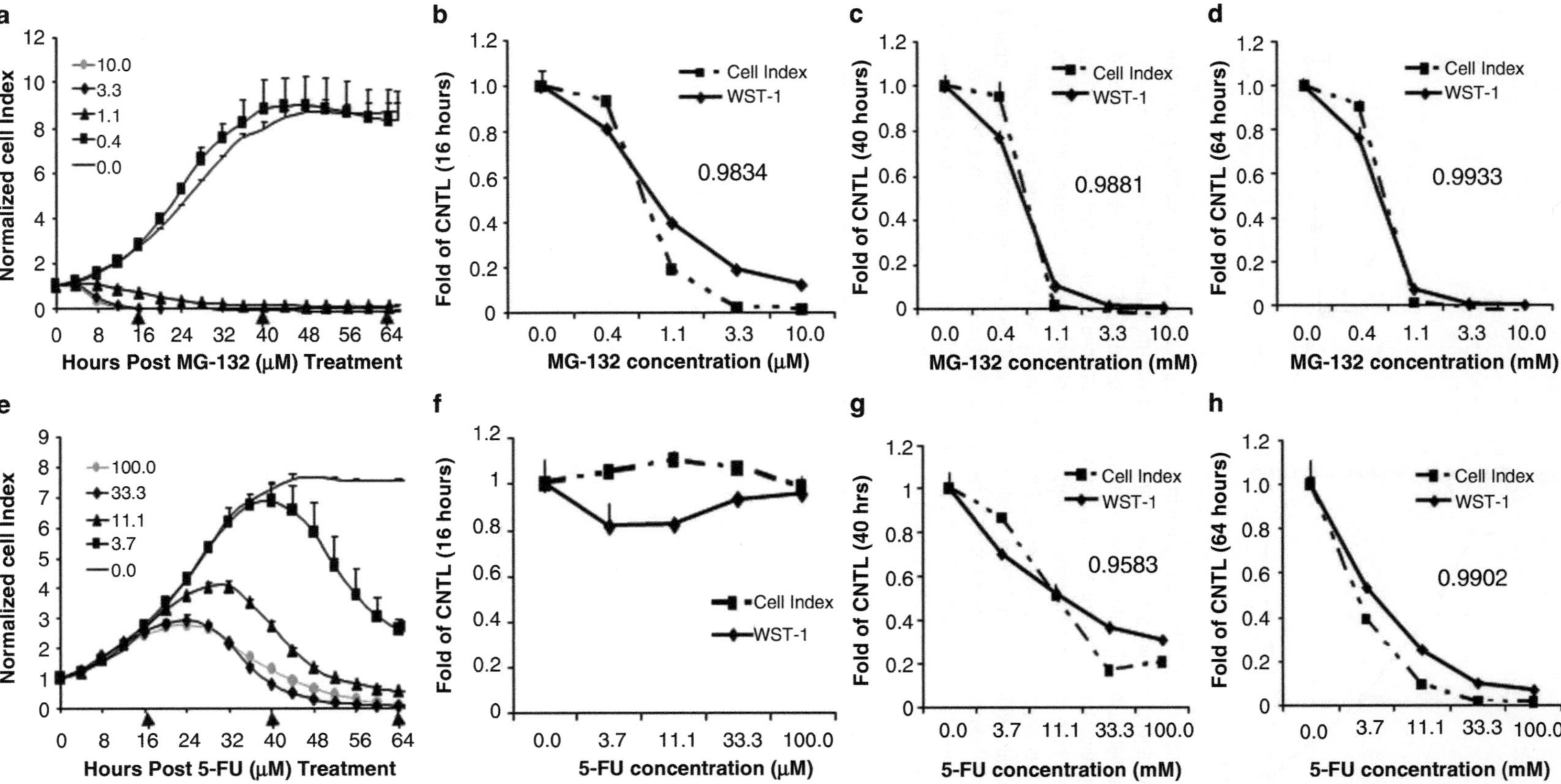

Fig. 2. xCELLigence system allows real-time monitoring of cell viability. HeLa cells were treated with various concentrations of the proteasome inhibitor MG-132 (**a–d**) and the DNA-damaging agent 5-FU (**e–h**). Cell index was monitored continuously for 64 h and the normalized cell index was derived by dividing the cell index value at each time point by the cell index at time of compound additions and plotted (**a** and **e**). No compound (no cpd) was used as control. Parallel experiments were also set up in E-plates for the WST-1 assay, which was performed at 16, 40, and 64 h post compound addition (**b–d** and **f–h**). Normalized cell index and WST-1 readings against control samples (untreated) were shown at the respective time points (16 h, **b** and **f**; 40 h, **c** and **g**; 64 h, **d** and **h**). The average of triplicate samples was shown and the error bars indicate standard deviation. The Pearson correlation coefficient was also shown for each case where cytotoxicity was observed.

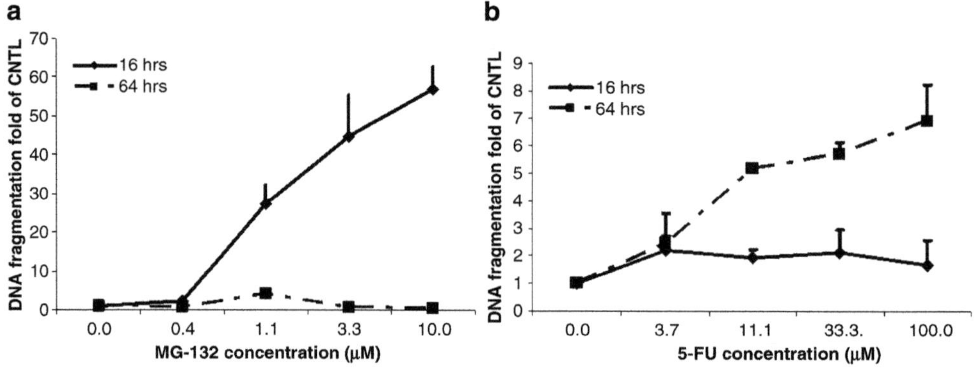

Fig. 3. Cell index changes and apoptosis. HeLa cells were treated with various concentrations of the proteasome inhibitor MG-132 (**a**) and the DNA-damaging agent 5-FU (**b**) and the cell index was monitored continuously for 64 h as in Fig. 2. A parallel experiment was set up in an E-plate and samples were harvested at 16 (*solid line*) and 64 h (*dotted line*) post compound addition for apoptosis assays (**a** and **b**). Apoptosis was determined using Cell Death Detection Elisa-Plus kit, and was expressed as the ratio of DNA fragmentation signal to that of the untreated samples. While the proteasome inhibitor MG-132 exerts its effect on cell index as early as 4 h post treatment, with maximal effect at 10 h post treatment, 5-FU effect is much slower, with onset at ~16 h post treatment, and the maximal effect observed after 48 h. Correspondingly, apoptosis induction is observed at 16 h for MG-132, and at 64 h for 5-FU. Importantly, no apoptosis induction is observed for MG-132 at 64 h, confirming the transient nature of apoptosis and indicating the importance of continuous monitoring of cell viability in selecting optimal time points for end-point assays.

2. Materials

2.1. Cell Culture and Compound Treatment

1. HyClone Dulbecco's Modified Eagle's Medium (DMEM) high Glucose is supplemented with 10% fetal bovine serum.
2. 0.05% Trypsin–EDTA.
3. Penicillin–streptomycin solution.
4. Hyclone DPBS/modified.
5. T25 and T75 cell culture flasks.
6. E-Plate 96 (Roche Applied Sciences, Indianapolis, IN).
7. CoStar 96-well cell culture plate.
8. MG-132 (Tocris, Ellisville, MO) is dissolved at 10 mM in DMSO, stored in –20°C, and then added to tissue culture as needed.
9. 5-Fluorouracil (5-FU, Sigma) is dissolved at 100 mM in DMSO, stored in –20°C, and then added to tissue culture as needed.

2.2. Cell Index Determination

1. xCELLigence and E-Plate 96 (Roche Applied Sciences, Indianapolis, IN).

2.3. WST-1 Assay

1. WST-1 reagent (Roche Applied Sciences, Indianapolis, IN). It is stored at –20°C until use.
2. DTX 800 plate reader (Beckman Coulter, Brea, CA).

2.4. Apoptosis Determination

1. Cell death Detection Elisa-Plus 96-tests kit (Roche Applied Sciences, Indianapolis, IN).
2. DTX 800 plate reader (Beckman Coulter, Brea, CA).

3. Methods

xCELLigence allows cell viability to be measured continuously in real time via impedance readout. Using two compounds with different cytotoxic kinetics (proteasome inhibitor MG-132 which takes ~12 h to observe the maximal cytotoxic effect (16), and the DNA damaging agent 5-FU which takes ~40 h) (17, 18), we have systematically documented the correlation of viability readouts between WST-1 and xCELLigence at several time points. The continuous monitoring of cell viability provides a time resolution of drug-mediated cytotoxicity and demonstrated the difference of time-dependent cellular response for the two compounds (Fig. 2). The continuous monitoring of cytotoxicity by xCELLigence system can be used to select the optimal time points (e.g., when the cell index is at the lowest level) to conduct more specific end-point assays (e.g., apoptosis assays) that are crucial to understand modes of cell death (Figs. 2 and 3).

3.1. Cell Culture and Compounds

1. HeLa cells were cultured in DMEM medium supplemented with 10% FBS and 1% penicillin–streptomycin (PS) in 37°C incubator with 5% CO_2. Cells were passaged when approaching confluence with 0.05% trypsin–EDTA and split at a ratio between 1:20 and 1:6.
2. Cells at 80–90% confluence were split into a T75 flask at 1:3 ratio the day before use.
3. On the day of seeding, the medium was removed, and cells were washed gently with 7 ml of PBS. 1 ml of trypsin was added, and the cells were incubated at 37°C for 1–10 min until cells lifted off the flask (see Note 1).
4. Add 9 ml of DMEM/10% FBS/1% PS to cell suspension. Pipette up and down to mix cells well, and transfer the cell suspension to a 50-ml conical tube.
5. Remove 10 μl of cell culture and count the cell number using a hemocytometer. In general, ~5×10^5 cells/ml is obtained.
6. Make 2.5×10^4 cells/ml working stock. 100 μl (2,500 cells) will be needed for each well.

3.2. Cell Index Determination

These instructions assume the use of Real-Time Cell Analyzer (RTCA) Single Plate (SP) instrument, although they are easily adaptable to RTCA-MP (Multiple Plate) or RTCA-DP (Dual Plate) instruments. Cell Index (CI) is an arbitrary unit reflecting the electronic cell-sensor impedance. It is defined as (Rn–Rb)/15 where Rn is the cell–electrode impedance of the well with the cells and Rb is the background impedance of the well with the medium alone. The cell index value directly correlates with the number of viable cells.

1. Start the RTCA software. It will automatically detect the RTCA instrument connected to RTCA control unit, and launch the appropriate user interface.
2. Fill out the experiment setup pages, including Experiment-note, Layout, and Schedule pages. In general, the schedule will include three steps: background measurement step; cell monitoring step; and compound activity monitoring step. In general, triplicate wells are used for each treatment for statistical analysis.
3. Add 50 μl of medium to each well of the E-Plate. Transfer the E-Plate to the desired SP instrument and start the experiment by measuring background (first step).
4. When background measurement is done, take out the E-Plate and put it inside the tissue culture hood. Add 100 μl of the cell suspension (2.5×10^4 cells/ml) to each well of the plate. Optimal cell number should be empirically determined for each cell type. Tap the plate gently ten times from each direction to mix well the cells. Leave the plate in the tissue culture hood at room temperature for 30 min.
5. Put the E-Plate in the instrument, and start the cell monitoring step (second step) (see Note 2).
6. 24 h later, thaw the compounds at room temperature. Dilute the compounds in DMEM medium so that the compound is 10× of the final concentration. DMSO is added as untreated control.
7. Stop the second step of the experiment and take out the plate. Add 16.7 μl of the 10× compound dilution to desired wells.
8. Put the E-Plate back to the instrument and start the compound activity monitoring step (third step). The length of monitoring will depend on the compounds and cell lines used. 48–72 h are generally used to document cytotoxic effects (see Note 3).

3.3. WST-1 Assay

For cell viability assay using WST-1, the seeding and treatment procedure are the same as described in the previous section (Subheading 3.2). At desired time points, the medium was removed from E-Plate, and replaced with appropriate medium containing 10% WST-1. Plates were incubated in a 37°C incubator with 5% CO_2 for 0.5–4 h, during which viable cells convert WST-1 to a water-soluble formazan dye that can be quantified by absorbance reading using a plate reader. The absorbance directly correlates with viable cell number.

1. At desired time points, remove the E-Plate from instrument in the incubator (see Note 4).
2. Remove the culture medium from wells of the E-Plate in the tissue culture hood.
3. Dilute WST-1 reagents 1:10 into culture growth medium. Add 120 μl to each well of the E-Plate, and place the Plate back into 37°C incubator with 5% CO_2.
4. 0.5–4 h later, remove the E-Plate from CO_2 incubator. Transfer 100 μl of medium/WST-1 from each well into a new 96-well tissue culture plate (see Note 5).
5. Measure absorbance at 450 nm using a plate reader. For this study, a Beckman DTX 800 Multi-Mode Detector is used.

3.4. Apoptosis Assay

For apoptosis, the Cell Death Detection Elisa-Plus kit is used. This is an Elisa-based kit that detects histone-complexed DNA fragments, which are indicative of apoptotic cells. The experimental setup is similar as described in Subheading 3.2. Cells were harvested at desired time points and assay conducted.

1. On the day of assay, prepare all the working solutions following manufacturer's instructions.
2. At the time of assay, remove the E-Plate from the instrument. Centrifuge the E-Plate at 200×*g* for 10 min to harvest both the floating and adherent cells.
3. Remove the culture medium completely. Add 200 μl lysis buffer to each well. Gently tap the E-Plate several times and incubate the plate at room temperature for 30 min.
4. Centrifuge the lysates at 1,100 rpm for 10 min. Transfer 20 μl of the supernatant into the streptavidin-coated plate (see Note 6). Please remember to include the positive control that comes with the assay kit in the assay.
5. Add 80 μl of the immunoreagent to each well containing samples.
6. Cover the microplate with an adhesive cover foil. Incubate on a shaker (300 rpm) for 2 h at room temperature.
7. Remove the solution completely, wash with 250 μl incubation buffer three times. Remove solution completely.
8. Add 100 μl ABTS solution. Incubate on a shaker (250 rpm) until the color development is sufficient for a photometric analysis (10–30 min) (see Note 7).
9. Add 100 μl ABTS stop solution.
10. Measure absorbance at 405 nm using a plate reader. Use ABTS+ABTS stop solution as blank. For the measurement, the Beckman DTX 800 Multi-Mode Detectors is used (see Note 8).

11. For calculation, the absorbance of samples is corrected for background. Fold increase of DNA fragmentation is calculated by dividing the absorbance of treated samples by the absorbance of nontreated sample.

4. Notes

1. Check cells frequently under the microscope to determine whether cells have been lifted off. Prolonged incubation with trypsin is toxic to the cells.
2. The continuous growth of the cell line is monitored in this step, and can serve as quality control for cell growth. In general, a measurement every 30–60 min should be sufficient.
3. This step is to monitor the compound-mediated effect on cell viability. In general, monitoring every 15 min for 48–72 h should be enough for most cytotoxic compounds. For compounds that mediate cellular morphological changes or induce immediate cytotoxicity, more frequent monitoring (e.g., every minute) is recommended for the first 1–2 h.
4. Alternatively, the WST-1 assay can also be set up and conducted in a regular 96-well tissue culture plate in parallel.
5. It is important to control the incubation time so that the reading is not saturated and a linear relationship is observed between the absorbance reading and cell number. It may be necessary to empirically determine the incubation time beforehand.
6. Gently remove 20 μl of the supernatant. Do not disturb the cell pellets, which contain high molecular weight, unfragmented DNA.
7. Positive control samples would turn dark green within 10 min. The time of incubation for samples of interest should be determined empirically.
8. If reading is saturated, e.g., OD 405 nm > 4.0, the samples should be diluted in the incubation buffer and measured again. Please note that all samples need to be diluted similarly for comparison.

References

1. Riss, T.L. & Moravec, R.A. (2004) Use of multiple assay endpoints to investigate the effects of incubation time, dose of toxin, and plating density in cell-based cytotoxicity assays. *Assay Drug Dev Technol* **2**, 51–62.
2. Mosmann, T. (1983) Rapid colorimetric assay for cellular growth and survival: application to proliferation and cytotoxicity assays. *J Immunol Methods* **65**, 55–63.
3. Scudiero, D.A. et al. (1988) Evaluation of a soluble tetrazolium/formazan assay for cell growth and drug sensitivity in culture using human and other tumor cell lines. *Cancer Res* **48**, 4827–33.

4. Berridge, M.V., Herst, P.M. & Tan, A.S. (2005) Tetrazolium dyes as tools in cell biology: new insights into their cellular reduction. *Biotechnol Annu Rev* **11**, 127–52.
5. Nociari, M.M., Shalev, A., Benias, P. & Russo, C. (1998) A novel one-step, highly sensitive fluorometric assay to evaluate cell-mediated cytotoxicity. *J Immunol Methods* **213**, 157–67.
6. de Fries, R. & Mitsuhashi, M. (1995) Quantification of mitogen induced human lymphocyte proliferation: comparison of alamarBlue assay to 3H-thymidine incorporation assay. *J Clin Lab Anal* **9**, 89–95.
7. Shahan, T.A., Siegel, P.D., Sorenson, W.G., Kuschner, W.G. & Lewis, D.M. (1994) A sensitive new bioassay for tumor necrosis factor. *J Immunol Methods* **175**, 181–7.
8. Petty, R.D., Sutherland, L.A., Hunter, E.M. & Cree, I.A. (1995) Comparison of MTT and ATP-based assays for the measurement of viable cell number. *J Biolumin Chemilumin* **10**, 29–34.
9. Koop, A. & Cobbold, P.H. (1993) Continuous bioluminescent monitoring of cytoplasmic ATP in single isolated rat hepatocytes during metabolic poisoning. *Biochem J* **295 (Pt 1)**, 165–70.
10. Crouch, S.P., Kozlowski, R., Slater, K.J. & Fletcher, J. (1993) The use of ATP bioluminescence as a measure of cell proliferation and cytotoxicity. *J Immunol Methods* **160**, 81–8.
11. Garewal, H.S., Ahmann, F.R., Schifman, R.B. & Celniker, A. (1986) ATP assay: ability to distinguish cytostatic from cytocidal anticancer drug effects. *J Natl Cancer Inst* 77, 1039–45.
12. Kroemer, G. et al. (2005) Classification of cell death: recommendations of the Nomenclature Committee on Cell Death. *Cell Death Differ* **12 Suppl 2**, 1463–7.
13. Ramirez-Ortega, M. et al. (2006) Proliferation and apoptosis of HeLa cells induced by in vitro stimulation with digitalis. *Eur J Pharmacol* **534**, 71–6.
14. Weinstein, I.B. & Joe, A. (2008) Oncogene addiction. *Cancer Res* **68**, 3077-80; discussion 3080.
15. Abassi, Y.A. et al. (2009) Kinetic cell-based morphological screening: prediction of mechanism of compound action and off-target effects. *Chem Biol* **16**, 712–23.
16. Han, Y.H., Moon, H.J., You, B.R. & Park, W.H. (2009) The effect of MG132, a proteasome inhibitor on HeLa cells in relation to cell growth, reactive oxygen species and GSH. *Oncol Rep* **22**, 215–21.
17. Yim, E.K. et al. (2004) Proteomic analysis of antiproliferative effects by treatment of 5-fluorouracil in cervical cancer cells. *DNA Cell Biol* **23**, 769–76.
18. Singh, S., Upadhyay, A.K., Ajay, A.K. & Bhat, M.K. (2007) Gadd45alpha does not modulate the carboplatin or 5-fluorouracil-induced apoptosis in human papillomavirus-positive cells. *J Cell Biochem* **100**, 1191–9.

Chapter 7

Analysis of Tumor and Endothelial Cell Viability and Survival Using Sulforhodamine B and Clonogenic Assays

Caroline Woolston and Stewart Martin

Abstract

A variety of assays, and rationales for their use, exist to monitor viability and/or survival following cellular exposure to insult. Two commonly used in vitro assays are the sulforhodamine B assay and the clonogenic survival assay which can be used to monitor the efficacy of anticancer agents, either via direct tumor cell cytotoxicity or antiangiogenic mechanisms. The techniques described are suitable for studying survival in a number of different cell types; however, this chapter describes how they may be used in the assessment of chemo-/radiosensitivity. The methods are uncomplicated and robust as long as attention is paid to key optimization steps. Except for a multiwell plate reader they do not require any specialized equipment other than that found in a typical tissue-culture laboratory.

Key words: Clonogenic assay, Sulforhodamine B, Cell survival, Tumor cell, Endothelial cell

1. Introduction

Measuring cell viability and survival after a toxic insult is a key process and one that is often misreported in the literature. There are numerous assays available that are used to assess cell killing/survival but without a combined understanding of the cell type being studied, or the treatment effects, these have the potential to severely under or over estimate the level of killing achieved (reviewed in (1, 2)). In certain cell types, such as leukemias, many lymphomas, and some solid tumors, apoptosis is the primary mode of cell death. After treatment, cells rapidly undergo apoptosis and show distinct morphological features such as altered membrane permeability, chromatin condensation/fragmentation, and exposure of phosphatidyl serine to the outer membrane. These changes can be measured using simple short-term assays such as trypan blue/propidium iodide exclusion or annexin V

Martin J. Stoddart (ed.), *Mammalian Cell Viability: Methods and Protocols*, Methods in Molecular Biology, vol. 740,
DOI 10.1007/978-1-61779-108-6_7, © Springer Science+Business Media, LLC 2011

staining. For other solid tumors, particularly carcinomas, cells do not immediately die after treatment, and apoptosis may not be the main mediator of cell killing. Therefore, short-term assays measuring early functional changes would provide no information on overall toxicity of treatment. It is important, therefore, to conduct assays that measure viability/survival in both the short- and long-term.

A true cell survival assay measures the end result of treatment which can either be cell death or recovery. The clonogenic assay has become the "gold standard" for assessing cellular sensitivity to cytotoxic treatments as it tests the fundamental aspect of survival, a cells ability to undergo sufficient proliferation so as to form a colony (2, 3). It is widely used for the assessment of radiosensitivity as it can accurately assess the long-term survival effects of irradiation. It is also used to assess the efficacy of novel anticancer drugs; however, the high-throughput nature of sulforhodamine B (SRB) and 3-(4,5-dimethylthiazol-2-yl)-2,5-diphenyltetrazolium bromide [MTT] assay makes them more amenable for drug discovery work. Each assay has inherent limitations (reviewed below) and caution should be used, as mentioned above, with interpretation of data if the cell type is less prone to early as opposed to late lethality, in which case the clonogenic assay should be utilized. The clonogenic assay is applicable for use with any type of cell that can be grown in culture, whether it is an animal, plant, yeast, or bacterial cell (4). Due to its very large dynamic range, it is good at very low and high survival extremes that limit other colorimetric and dye-exclusion-based assays.

When comparing SRB and clonogenic assays, the latter is more laborious and time-consuming (both in terms of conducting the assay and in generation of results, i.e., plates have to be incubated for 2–4 weeks), and problems can ensue with cell lines that do not readily dissociate into single cell suspension. Another inconvenience, for certain labs, may be that a separate incubator is recommended for incubating plates to allow colony formation. Caution should be used in interpretation of data from cell types that have a low colony-forming ability (plating efficiency) as results may not be representative of the population as a whole. The conventional (2D) clonogenic assay does not take into account the influence of intercellular communications that, in certain circumstances, play a key role in the cellular response to particular agents, such as those that induce DNA damage (5) – a number of novel, 3D-based, cultures and assays are in development but are not in routine use and are beyond the scope of this chapter.

The SRB assay is a popular in vitro cytotoxicity assay developed by Skehan and colleagues (6). SRB is an aminoxanthene dye that binds to basic amino acid residues in proteins through its sulfonic groups. The binding of SRB is stoichiometric; therefore, the amount of colorimetric change gives an estimation of total protein mass, which is directly proportional to cell number (7). Caution should be used as there will be certain instances where cell mass increases without a corresponding increase in cell number. For example, inhibition or alteration of the cell cycle process through drug treatment, or changes in growth conditions, is a known contributor to hypertrophy in certain cell types (8, 9). As mentioned above, the main application of the SRB assay is in chemosensitivity testing as it is particularly adaptable and suitable for high-throughput systems such as drug screening. For radiosensitivity testing, colorimetric assays are often thought to be inadequate due to the short duration of the assay; however it has been reported, for the SRB assay, that if sufficient time is left to allow delayed radiation-induced cell death (i.e., left for the equivalent of six cell doubling times or the equivalent length of time that it would take for the cells to form colonies of 50+ cells in a clonogenic assay) then comparable survival curves can be achieved (10). Caution needs to be used before application of this assay for radiosensitivity measurements as there is still some concern that at certain doses [i.e., high (>6 Gy) or low (≤2 Gy)] it is insensitive, i.e., it under or over estimates the survival compared to the clonogenic assay (11, 12).

In comparison to MTT, SRB staining is not dependent on mitochondrial function and so less variation is observed between cell lines leading to less cell-line specific optimization (10). SRB has better linearity and higher sensitivity than other colorimetric assays, such as Lowry, Bradford, azure A, and thionin (6), and is comparable to those achieved with standard fluorescent dye-staining methods, such as DAPI and Hoechst 33342 (7). The background staining due to fixation of medium components (fetal bovine serum) and dislodgment of adherent cells can be improved with optimization giving good signal:noise ratios and CV values (13). From a practical perspective, it does not require any specialized equipment apart from a multiwell plate reader (see Subheading 2), has a stable endpoint that can be stored indefinitely, akin to the clonogenic assay, and is therefore overall very cost-effective. However, the multiple washing and drying steps discourage some people using this assay. The TCA addition must be carefully optimized to minimize the risk of loss of cells during fixation and, as with the clonogenic assay, care must be taken in the plating of cells to avoid artifactual results due to uneven plating. Although the SRB is widely used for tumor cell chemosensitivity testing, it is less sensitive with endothelial cells due to their relatively lower protein levels (14).

If characterizing novel agents (radio- or chemotherapy), the choice of assay will largely depend on the cell type being studied but using both the SRB and clonogenic assays, at least for an initial comparison, is recommended.

2. Materials

2.1. Clonogenic Assay

1. Appropriate cell culture medium.
2. Phosphate-buffered saline (PBS).
3. Trypsin/EDTA.
4. 15/50 ml sterile tubes.
5. T25 tissue culture flasks or 100 mm cell culture treated dishes or six-well plates (see Note 1).
6. Dedicated clonogenic humidified incubator set at 37°C with an atmosphere of 5–10% CO_2 depending on the cell type.
7. Cell counting system (e.g., hemocytometer, Coulter counter, Invitrogen "Countess" system).
8. Saline (0.9% w/v).
9. Methanol.
10. 0.5% crystal violet in water (Sigma).
11. Appropriate colony counting device – marker pen, colony counting pen, automated system (see Subheading 3).
12. Inverted phase contrast microscope.

2.2. Sulforhodamine B Assay

1. Appropriate cell culture medium.
2. PBS.
3. Trypsin/EDTA.
4. 15/50 ml sterile tubes.
5. 96-Well flat bottomed polystyrene tissue culture plates.
6. Cell counting system (e.g., hemocytometer, Coulter counter, Invitrogen "Countess" system).
7. Dimethyl sulfoxide (DMSO).
8. Ice-cold 50% (w/v) trichloroacetic acid (TCA).
9. 1% (v/v) acetic acid.
10. 0.4% (w/v) SRB (Sigma) in 1% (v/v) acetic acid.
11. 10 mM unbuffered Tris base solution.
12. Multiwell microplate reader (absorbance – 490–530 and 620 nm. Fluorescence – excitation 488 nm and emission 585 nm).
13. Gyratory plate shaker (optional).
14. Inverted phase contrast microscope.
15. Humidified incubator set at 37°C with an atmosphere of 5–10% CO_2 depending on the cell type.
16. Refrigerator set at 4°C.

3. Methods

3.1. Clonogenic Assay

The method outlined below is for adherent cell lines. Suspension cells can also be assessed via clonogenic assay using the adaptations outlined in Subheading 3.1.4.

3.1.1. Seeding Approaches

There are two different approaches to the seeding of cells that will subsequently undergo clonogenic incubation (see Note 2). The first approach is to harvest cells from logarithmically growing stock cultures and plate them at the required final clonogenic dilutions prior to treatment. Plates are initially incubated for long enough to allow cell attachment, but treatment must be applied before they have a chance to start proliferating, increasing the number of potential colonies, and giving a false result. The initial attachment period must be optimized for each individual cell line. After treatment, the dishes are placed in an incubator for a time equivalent to at least six potential cell divisions (to give colonies of >50 cells). This method is often used for a quick screening of the sensitivity of cells to different treatments.

The second, and more common, approach is to treat logarithmically growing cells and subsequently replate for clonogenic assessment (substantial differences are observed, for certain agents, if plateau phase cells are treated). The replating may be performed immediately after treatment or may be delayed to examine for the effect of repair processes.

Irrespective of the method chosen, cells are generally plated in triplicate and, to account for potential interexperimental differences in plating efficiencies, at two-cell densities (see Note 3).

3.1.2. Plating Before Treatment

1. Harvest exponentially growing cells as stated in Subheading 3.1.1.
2. Count cells and plate an appropriate number of cells per dish/flask (see Notes 4 and 5).
3. Allow time for the cells to adhere to the plastic (see Note 6).
4. Treat the cells as required for the experiment being careful not to disturb the cells (see Note 7).
5. Place the cells in the incubator and leave until control dishes have formed sufficiently large clones (see Notes 8–10). *Reminder*: Experimental plates should NOT be disturbed before the end of the appropriate incubation time as this can cause cells to detach from initial colonies and form new colonies that will skew results.

3.1.3. Plating After Treatment

1. Harvest the cells quickly after treatment to stop any erroneous results occurring due to cellular repair (see Note 11).

2. Cell count and plate an appropriate number of cells based on treatment severity to achieve the optimal number of colonies (see Notes 3, 4, and 12).
3. Place the cells in the incubator and leave until control dishes have formed sufficiently large clones (see Notes 8–10). *Reminder*: Experimental plates should NOT be disturbed before the end of the appropriate incubation time as this can cause cells to detach from initial colonies and form new colonies that will skew results.

3.1.4. Suspension Cells and Difficult Cell Types

Some cells when plated at low density do not form colonies and therefore require additional steps to help ensure a good plating efficiency. Human umbilical vein endothelial cells (HUVEC) are routinely cultured on 0.2% gelatin and will form colonies as long as flasks/dishes are also coated with gelatin – if the gelatin coating is omitted then the plating efficiency can suffer. Three other alternative methods to enable successful colony formation and optimal plating efficiencies are use of soft agar (especially for suspension cells), incubating with conditioned media, or use a feeder layer. The feeder layer may be a layer of cells such as fibroblasts that have been irradiated to not divide anymore but still produce growth-promoting factors.

3.1.5. Fixation and Staining of Colonies

1. Very carefully aspirate the medium from the plates/flasks so as not to disturb the colonies (see Note 13).
2. Gently wash the plates/flasks with PBS.
3. Add methanol:saline (1:1) in a sufficient volume to cover the colonies. Leave for 15 min.
4. Remove and add 100% methanol for a further 15 min.
5. Remove and add 0.5% crystal violet solution and leave to allow sufficient staining (see Notes 14 and 15).
6. Remove the crystal violet solution (N.B. this can be reused) and carefully rinse with an indirect flow of tap water.
7. Invert and leave to dry on the bench, at room temperature.

3.1.6. Colony Counting Approaches

There are numerous ways in which the colonies can then be counted. The simplest way is to take a pen and manually mark and count each colony on the underside of the plate or to use a "colony counter pen."

Analysis machines are also available to enable automated counting of colonies such as the gelcount system from Oxford Optronix. These systems can give additional information on colony size, statistical parameters such as colony OD and theoretical colony "volume" alongside the higher throughput of plates, and objective and consistent counting.

3.1.7. Calculating Plating Efficiency and Survival Fraction

The plating efficiency (PE) is the number of colonies formed under control conditions, i.e., it is the ratio of the number of colonies to the number of cells seeded.

The PE is calculated from the untreated controls using the following equation:

$$PE = \frac{\text{no. of colonies formed}}{\text{no. of cells seeded}} \times 100\%$$

The surviving fraction (SF) is the number of colonies that form after the cells have been treated:

$$SF = \frac{\text{no. of colonies formed after treatment}}{\text{no. of cells seeded} \times PE}$$

For radiation analysis, survival curves are usually plotted with SF as a log scale and the treatment dose as a linear scale. A variety of parameters can be assessed from such radiation dose survival plots to gauge parameters such as inherent sensitivity, repair capacity, etc. For example the mean lethal dose (37% of the original starting number) D_{37} or D_0 value can be calculated from the linear, exponential, portion of the curve.

For chemosensitivity, the graph of survival can either be plotted on a linear scale or, if the range of treatments is large, on a log scale. From this, if a comprehensive dose range has been conducted, the dose at which 50% cell death is achieved can be calculated (IC50).

3.2. Sulforhodamine B Colorimetric Assay

3.2.1. Cell Plating

1. Seed cells into a 96-well plate at optimal density in 180 μl. This can range from 1,000 to 40,000 cells/well depending on the cell line being used, the treatment and the assay time course. On a separate plate, at least three wells must be set up for a Day 0 estimation of cell population at time of treatment (see Notes 5 and 16).
2. Incubate the plates at optimal growth conditions, i.e., 37°C, 5–10% CO_2 for 12–24 h.
3. Process wells set aside for Day 0 assessment – see Subheading 3.2.2.
4. Add required drug treatments to each well in 20 μl volumes so the final well volume equals 200 μl. Incubate cells for the required time then proceed to cell fixation and staining.

3.2.2. Cell Fixation and Staining

The fixation procedure can be conducted in different ways depending on whether the cell type being used is adherent (1a), semi-adherent, or suspension (1b).

1. (a) For adherent cells. Aspirate media and add 200 μl of PBS to each well. Gently add 50 μl of ice-cold 50% stock of TCA to each well, creating a final concentration of 10% TCA. TCA concentration can be optimized for different cell lines (see Notes 17 and 18). (b) For semi-adherent or suspension cells. Gently add 50 μl of ice-cold 80% stock of TCA to each well containing media making sure not to disturb the settled cells (see Notes 17–19).
2. Place the plate at 4°C for 1 h to allow cell fixation (see Note 18).
3. Discard supernatant and gently wash the plates 4–5 times with tap water and air dry being careful not to disturb the fixed cells. *At this stage the plates can be stored indefinitely at room temperature.*
4. Add 100 μl of SRB solution [0.4% (w/v) SRB in 1% acetic acid] to each well and incubate for 10 min at room temperature. SRB concentration can be optimized for different cell lines (see Note 14).
5. Wash plates 4–5 times with 1% acetic acid to remove unbound dye, invert, and air dry at room temperature. *At this stage the plates can be stored indefinitely at room temperature* (see Note 20).
6. Add 200 μl of 10 mM Tris base solution (pH 10.5) to each well to solubilize the dye. The plate can either be placed on a gyrator shaker for 5 min or if not available, left at room temperature for 30 min with occasional shaking (see Note 21).
7. The plate absorbance should be read between 490 and 530 nm subtracting the background measurement at 620 nm. Alternatively, the SRB can be measured fluorometrically at an excitation of 488 nm and emission of 585 nm.

3.2.3. Calculation of Percentage Cell Growth Inhibition

To calculate the percentage of cell growth inhibition, the following calculations can be used.

If the mean OD of the sample is greater than or equal to the mean OD of the Day 0 sample:

1. $$\%\text{ of control cell growth} = \frac{\text{meanOD}_{\text{sample}} - \text{meanOD}_{\text{day 0}}}{\text{meanOD}_{\text{negcontrol}} - \text{meanOD}_{\text{day 0}}} \times 100$$

If the mean OD of the sample is less than the mean OD of the Day 0 sample:

2. $$\%\text{ of control cell growth} = \frac{\text{meanOD}_{\text{sample}} - \text{meanOD}_{\text{day 0}}}{\text{meanOD}_{\text{day 0}}} \times 100$$

% growth inhibition = 100 – % of control cell growth

Growth inhibition of 50% (GI50) is demonstrated when calculation (1) = 50. The drug concentration that results in total growth inhibition (TGI) is obtained when the mean OD of the sample = mean OD at Day 0. The concentration of drug resulting in a 50% reduction at the end of the drug treatment as compared to that at the beginning (LC50), indicating a net loss of cells, is obtained when calculation (2) = –50.

4. Notes

1. Optimization of tissue culture plate/flask suppliers may be required to achieve optimal plating efficiencies.
2. The choice of seeding approach will depend on the type and severity of the treatment, e.g., if high doses of radiation are being studied. Initial plating should be at a high enough density, but ensuring logarithmic growth, to recover enough cells for analysis. Numerous plates may be required so that the cells are not confluent prior to treatment. If cells are initially plated at too high a density then it is also difficult, if plating for colonies pre-treatment, to guarantee that colonies have been formed from one cell rather than two sitting next to each other. In these circumstances, most investigators prefer seeding post treatment.
3. As a guide, the number of colonies that would be optimal to achieve at the end of a clonogenic assay on each plate would be 100 for a P100 dish (55 cm^2) and 50 for a T25 flask (25 cm^2). This would give an average colony per area of two colonies per square centimeter.
4. Accurate cell counting is essential to enable the correct number of cells to be plated. This will enable an accurate plating efficiency (PE) to be determined from the untreated controls and for the final survival calculations for the treated groups. The number of cells that will need to be plated will depend on the severity of the treatment being conducted. To begin with, use different dilutions of control cells and those at the maximum treatment dose to determine the most appropriate seeding density.
5. Appropriate controls should also be included, i.e., untreated cells, treatment with a drug that is known to have an effect (positive control 1) and/or a control cell line that the outcome of the treatment of interest is known (positive control 2).
6. An initial time course should be set up to establish the optimal time to adherence, prior to treatment, for each cell type being used.

7. To avoid disturbing the cells and potentially removing a susceptible population, it is advised to add the treatment to the existing media rather than replacing via aspiration or pipetting.
8. Sufficient colonies are generally termed those that contain in excess of 50 cells. Colonies that contain less than 50 cells should not be counted as they have lost reproductive integrity. It is a good idea to determine the optimal time that the control colonies require to form sufficiently large enough colonies before starting any experiments. This can be done by plating a range of cell dilutions for multiple time points, such as 2, 3, or 4 weeks. At each time point, plates can be carefully removed and stained to assess colony size.
9. Ideally a separate incubator is used for the duration of the clonogenic assay to prevent colonies being disturbed during colony formation. Disturbance of the colonies, particularly shortly after the start of the incubation period, will cause cells to detach from initial colonies and form satellite colonies, giving artifactual increased colony numbers.
10. The incubator conditions need to be optimal for the growth conditions of the cell line being used, i.e., if the cells are normally maintained at 37°C, 5% CO_2, then these are the conditions that should be adhered too. It is also important that the incubator being used has a humidity tray to prevent drying/evaporation of the culture medium, especially if dishes are being used, as some cultures may be left for up to 4 weeks.
11. Minimize the amount of time the cells are handled before plating to prevent changes in pH, temperature, and other external influences altering the cells viability. Prepare all tissue culture equipment and reagents prior to commencing.
12. For cell counting after treatment, trypan blue should not be used as a total cell count is required, i.e., the total number of alive and dead cells. If the dead cells are excluded in the count, this will cause false results.
13. Following incubation, carefully remove the plates from the incubator and treat with caution until the colonies are fixed.
14. The concentration and length of incubation of the crystal violet solution or SRB dye used will depend on the cell line. Optimize the concentration of the dye used first as prolonged time in the solution may cause lifting and thus loss of the colonies/protein.
15. There are different fixing and staining methods and another common method is to wash with PBS and then add 6% glutaraldehyde with 0.5% crystal violet. Again, depending on the cell type that is being used, the stain concentration and length of incubation may have to be optimized.

16. Ensure homogeneous plating in each well is achieved as this is a common cause of error in these assays. During the plating process make sure the cells to be plated are regularly mixed to avoid problems.
17. TCA can fix both cells and media proteins so removal of the media results in lower background levels enhancing the sensitivity and reliability of the assay especially at low cell numbers.
18. The TCA concentration can be optimized for individual cell types. Loosely attached cells may require a higher TCA concentration and extended incubation at 4°C, whereas TCA concentrations can be reduced for fully adherent lines. TCA fixation is critical for a successful assay.
19. Plates containing suspension or semi-adherent cells may be centrifuged if a plate rotor is available prior to addition of the TCA reagent. Care must still be taken in the addition of TCA to prevent disturbance of the settled layer.
20. Insufficient removal of the SRB dye can cause an over-estimation of the cell numbers present, however excessive, prolonged washing has the potential to bleach the protein-bound dye or disrupt the fixed layer. It is advised to rinse several times but quickly.
21. The pH of the solutions used is critical for the SRB dye steps. Binding of the SRB occurs under mild acidic conditions but is dissociated under basic conditions.

References

1. Brown, J.M. and Wouters, B.G. (2001) Apoptosis: mediator or mode of cell killing by anticancer agents? *Drug Resis Updates* **4**, 135–6.
2. Brown, J.M. and Wouters, B.G. (1999) Apoptosis, p53, and tumor cell sensitivity to anticancer agents. *Cancer Research* **59**, 1391–9.
3. Franken, N.A.P., Rodermond, H.M., Stap, J., Haveman, J., van Bree, C. (2006) Clonogenic assay of cells *in vitro. Nature Protocols* **1** (5), 2315–19.
4. Josephy, P.D. and Mannervik, B. Molecular Toxicology. 2nd ed. Oxford University Press, 2006.
5. Garvey, R.B. New research on cell aging. Nova Science Publishers Inc, 2007.
6. Skehan, P., (1990) New colorimetric cytotoxicity assay for anticancer-drug screening. *J. Natl. Cancer Inst* **82**, 1107–12.
7. Vichai, V. and Kirtikara, K. (2006) Sulforhodamine B colorimetric assay for cytotoxicity screening. *Nature Protocols* **1** (3), 1112–6.
8. Terada, Y., Inoshita S., Nakashima, O., Tamamori, M., Ito, H., Kuwahara, M., Sasaki, S., Marumo F. (1999) Cell cycle inhibitors ($p27^{Kip1}$ and $p21^{CIP1}$) cause hypertrophy in LLC-PK1 cells. Kidney Int **56**, 494–501.
9. Zhou, L., Li, J., Goldsmith, A.M., Newcomb, D.C., Giannola, D.M., Vosk, R.G., Eves, E.M., Rosner, M.R., Solway, J., Hershenson, M.B. (2004) Human bronchial smooth muscle cell lines show a hypertrophic phenotype typical of severe asthma. Am J Respir Crit Care Med **169**, 703–711.
10. Pauwels, B., Korst, A.E.C., de Pooter, C.M.J., Pattyn, G.G.O., Lambrechts, H.A.J., Baay, M.F.D., Lardon, F. and Vermorken, J.B. (2003) Comparison of the sulforhodamine B assay and the clonogenic assay for in vitro chemoradiation studies. *Cancer Chemother Pharmacol* **51**, 221–6.
11. Griffon, G., Merlin, J.L. and Marchal, C. (1995) Comparison of sulforhodamine B, tetrazolium and clonogenic assays for in vitro

radiosensitivity testing in human ovarian cell lines. *Anticancer Drugs* **6**, 115–23.

12. Banasiak, D., Barnetson, A.R., Odell, R.A., Mameghan, H. and Russell, P.J. (1999) Comparison between clonogenic, MTT and SRB assays for determining radiosensitivity in a panel of human bladder cancer cell lines and a ureteral cell line. *Radiat Oncol Invest* 7, 77–85.
13. Papazisis, K.T., Geromichalos, G.D., Dimitriadis, K.A. and Kortsaris, A.H. (1997) Optimization of the sulforhodamine B colorimetric assay. *J Immunol Methods* **208**, 151–8.
14. Connolly, D.T., Knight, M.B., Harakas, N.K., Wittwer, A.J., Feder, J. (1986) Determination of the number of endothelial cells in culture using an acid phosphatase assay. *Anal Biochem* **152,** 136–40.

Chapter 8

Annexin V/7-AAD Staining in Keratinocytes

Maya Zimmermann and Norbert Meyer

Abstract

Annexin V/7-amino-actinomycin staining is a convenient way to discriminate early apoptosis from late apoptosis and necrosis. Early apoptotic cells express phosphatidylserines (PS) on the outer leaflet of the plasma membrane. PS can be stained by labeled annexin V. Late apoptotic cells and necrotic cells lose their cell membrane integrity and are permeable to vital dyes such as 7-AAD (DNA intercalator).

Key words: Apoptosis, Annexin V, 7-AAD, Vital dye, DNA intercalator, Phosphatidylserine

1. Introduction

Annexin V/7-amino-actinomycin (7-AAD) staining is a method to discriminate between early apoptosis and late apoptosis and necrosis. The word apoptosis has been introduced in 1972 to describe a cell death with specific morphological features (1). The term apoptosis is a morphological description of characteristic aspects including rounding-up of the cell, plasma membrane blebbing, reduction of cellular volume (pyknosis), condensation of the chromatin, fragmentation of the nucleus (karyorhexis), and loss of plasma membrane asymmetry. Necrosis is often defined in a negative manner, as cell death lacking the characteristics of apoptosis and autophagy processes (2). The morphological appearance is usually similar to that of oncosis meaning cytoplasmic swelling, mechanical rupture of the plasma membrane, dilation of cytoplasmic organelles (mitochondria, endoplasmic reticulum, and Golgi apparatus), and moderate chromatin condensation.

Viable cells have an asymmetric distribution of different phospholipids between the inner and outer leaflets of the plasma membrane maintained by flippases. Phosphatidylserine (PS) is a

Martin J. Stoddart (ed.), *Mammalian Cell Viability: Methods and Protocols*, Methods in Molecular Biology, vol. 740,
DOI 10.1007/978-1-61779-108-6_8, © Springer Science+Business Media, LLC 2011

phospholipid component of the cell membrane and usually exposed on the cytosolic side of the membrane. When a cell undergoes apoptotic cell death, PS becomes exposed on the outer leaflet of the plasma membrane. Macrophages recognize PS on the cell surfaces and subsequently engulf and phagocyte these cells before membrane integrity is lost.

Annexin V has a high affinity to PS and binds therefore to cells with exposed PS. Since annexin V can be conjugated to fluorochromes it serves as a sensitive probe for flow cytometric analysis of cells undergoing apoptosis. 7-AAD is a DNA intercalator and used as a vital dye. As long as the cell keeps its membrane integrity, 7-AAD cannot bind/intercalate to the DNA since its penetration through the membrane is not possible. However, when the membrane integrity is lost, 7-AAD becomes access to the DNA cells (Fig. 1).

Therefore, viable cells are annexin V and 7-AAD negative, early apoptotic cells are annexin V positive and 7-AAD negative, and late apoptotic and necrotic cells are annexin V and 7-AAD positive. Annexin V/7-AAD staining does not distinguish between cells that have undergone apoptotic death versus those that have died as a result of a necrotic pathway since in either case dead cells will be stained by both labeled annexin V and 7-AAD.

Finally, it should be noted that PS exposure can occur on cells for phagocytic removal independent of apoptosis (3) and that PS exposure can be compromised in cells in which autophagy is impaired (4).

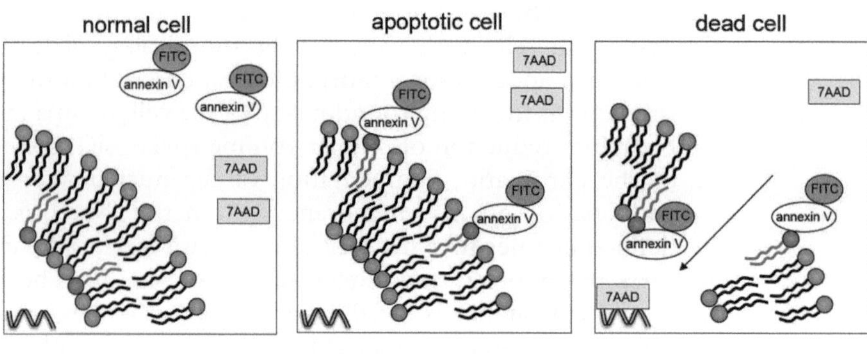

Legend:

phoshpholipids

phosphatidylserine (PS)

DNA

Fig. 1. Viability and apoptosis detection by annexin V and 7-AAD staining. Annexin V binds to phosphatidylserine (PS) and 7-AAD intercalates with DNA. Neither annexin V nor 7-AAD can penetrate through an intact cell membrane. In normal cell, PS exists only in the inner face of cell membranes. During apoptotic cell death, PS gets exposed at the outside of the membrane. Therefore, viable cells are annexin V and 7-AAD negative, apoptotic cells are annexin V positive, and 7-AAD negative and late apoptotic and necrotic cells are annexin V and 7-AAD positive.

2. Materials

2.1. Keratinocyte Culture

1. Keratinocyte-SFM medium (Invitrogen, San Diego, CA) supplemented with 5 ng/ml human recombinant epidermal growth factor (Invitrogen), 50 µg/ml bovine pituitary extract (Invitrogen), and 2 µM l-glutamine (Invitrogen).
2. Trypsin (0.25%) EDTA solution (Sigma–Aldrich, St. Louis, MO).
3. Trypsin neutralizing solution (Invitrogen).
4. Cyclohexamide (CHX) (Sigma–Aldrich).
5. Primary keratinocytes (Invitrogen) in passages 3–5.
6. T-75 Tissue Culture Flask (Milian AG, Basel, Switzerland).
7. 24-Well plates (BD, Franklin Lakes, NJ).

2.2. Annexin V and 7-AAD Staining

1. Annexin V-FITC solution (Beckmann Coulter, Nyon, Switzerland), concentration 2.5 µg/ml annexin V-FITC.
2. 7-AAD solution 50 mg/ml (Beckmann Coulter).
3. 10× Concentrated binding buffer (Beckmann Coulter) diluted 1:10 in distilled water.
4. 2 ml tubes and FACS analysis tubes.
5. Flow cytometer and software (EPICS XL-MCL, Expo32 software, Beckman Coulter).

3. Methods

3.1. Keratinocyte Culture and Splitting

1. Take the vial with your keratinocytes out of the nitrogen storage and thaw the cells in a 37°C water-bath until only small pieces of ice remain.
2. Transfer keratinocytes in a 1,000-µl pipette-tip to a T-75 Tissue Culture Flask with 40 ml 37°C preheated keratinocytes-SFM medium. Push the keratinocytes very slowly out of the pipette-tip (one drop per second). Be very careful here! (see Note 1).
3. Place the T-75 Tissue Culture Flask with the keratinocytes in an incubator with humidified atmosphere containing 5% CO_2 at 37°C. Let the cells adhere to the bottom and proliferate. Change medium every second day.
4. When keratinocytes reach a confluence of 70–80%, discard supernatant, add 5 ml preheated trypsin solution, and place the T-75 Tissue Culture Flask for 3–5 min in the incubator at 37°C. Keratinocytes will detach from the bottom (see Note 2).

5. Take the T-75 Tissue Culture Flask out of the incubator and add 10 ml preheated trypsin neutralizing solution to stop trypsin reaction.
6. Transfer all volume from the T-75 Tissue Culture Flask and centrifuge for 5 min at 350 × *g*.
7. Count the keratinocytes and transfer 3,000–5,000 cells per well to a 24-well plate. Place the 24-well plate in the incubator and let the cell settle and proliferate (see Note 3).
8. Stimulate keratinocytes with your stimulus so that cell confluence is less than 100% at the harvesting day. Use as negative control untreated keratinocytes and as positive controls keratinocytes stimulated with 10 and 100 ng/ml CHX for 4 days (see Note 4).

3.2. Annexin V and 7-AAD Staining Procedure

1. Collect supernatants of keratinocyte cultures and store on ice in order to collect the nonviable cells that have detached from the plastic.
2. Trypsinize keratinocytes grown in 24-well plates with 200 μl trypsin solution for 5 min at 37°C in the incubator (see Notes 2 and 5).
3. Add 400 μl of trypsin neutralizing solution to each well to stop trypsin reaction.
4. Transfer all volume from the wells to 2 ml tubes with the collected supernatants and centrifuge for 5 min at 500 × *g* at 4°C (see Note 6).
5. Discard supernatants and resuspend cell pellets in ice-cold 1× binding buffer to 5×10^6–10×10^6 cells/ml.
6. Tubes should be kept on ice to arrest further progress of the cells through the stages of viability, apoptosis, and necrosis.
7. Add 20 μl of 7-AAD dye solution and 10 μl annexin V-FITC solution to 100 μl of the cell suspension and mix gently.
8. Incubate tubes on ice for 15 min in the dark.
9. Add 400 μl of ice-cold 1× binding buffer and mix gently.
10. Analyze cell preparations within 30 min by flow cytometry. Store cells on ice (see Note 7).

3.3. Flow Cytometry

1. Use logarithmic scale for FL1 (525 nm), FL4 (675 nm) and sideward scatter (SS) and linear scale for forward scatter (FS).
2. Create biparametric histogram with SS LOG versus FS LIN and gate keratinocytes. For this you have to exclude small cell particles of dead cells, which are recognized as a tail in the lower part of the dot plot diagram (Fig. 2a). Use this gate for the following apoptosis analyses.
3. Set the flow cytometer such that the distribution of the annexin V-FITC negative population is in the first two decades

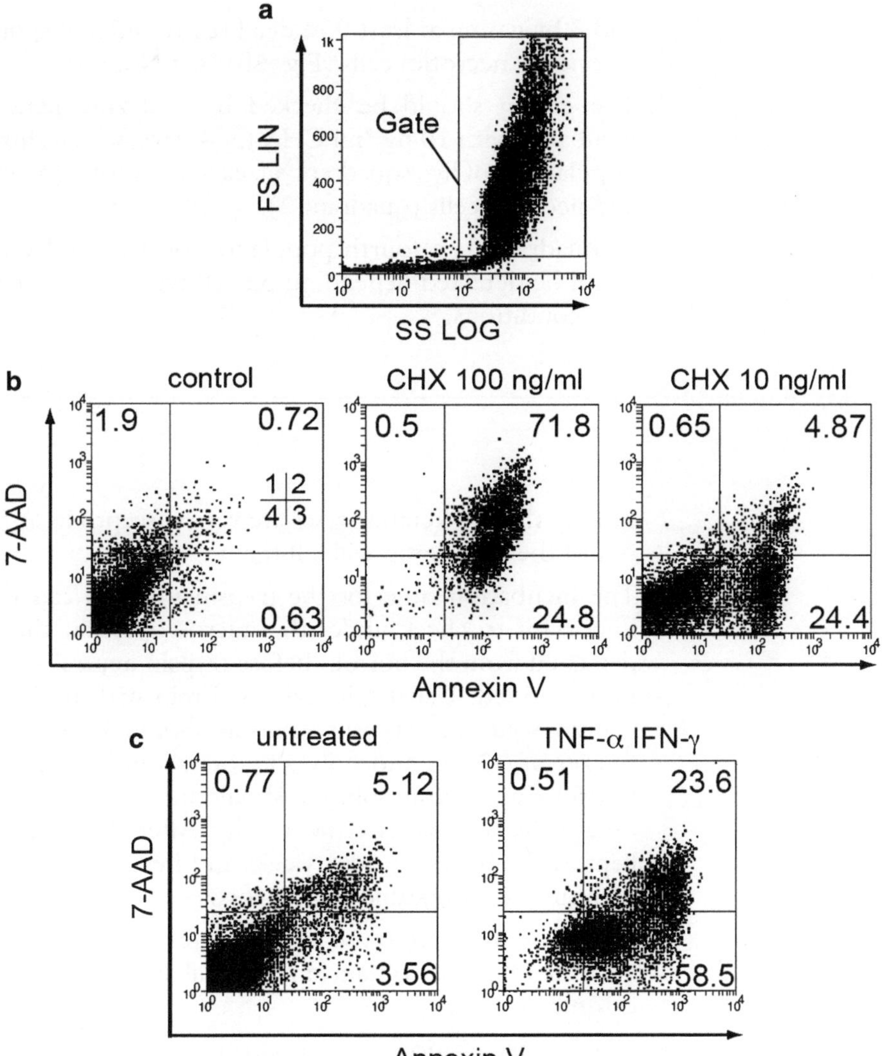

Fig. 2. Gating strategies for apoptosis analysis. Keratinocytes are gated to exclude small cell particles for the apoptosis analysis, which are recognized as a tail in the lower part of the *dot plot* diagram (**a**). To set quadrants as statistic markers, untreated keratinocytes were used as live cells (**b**, *left chart*) and for dead cells, keratinocytes were incubated with 100 ng/ml CHX (**b**, *middle chart*) for 4 days. The quadrants were set by gating at least 95% viable cells (quadrant 4) and at least 95% dead cells (quadrants 2 and 3). When keratinocytes were incubated with 10 ng/ml CHX for 4 days (**b**, *right chart*) the *dot plot chart* shows three different populations: (1) viable cells with low Annexin V and 7-AAD signal (quadrant 4), (2) apoptotic cells with high annexin V and low 7-AAD signal (quadrant 3), (3) necrotic cells with high annexin V and high 7-AAD signal (quadrant 2). Example data of annexin V and 7-AAD staining of untreated keratinocytes and keratinocytes incubated with TNF-α (10 ng/ml) and IFN-γ (10 ng/ml) for 4 days are presented (**c**).

of the FL1 parameter and the distribution of the 7-AAD negative population is in the first two decades of the FL4 parameter.

4. Untreated keratinocytes are used to set the fourth quadrant by gating at least 95% viable cells and keratinocytes treated with 100 ng/ml CHX for 4 days are used to set quadrants 2

and 3 by gating at least 95% dead (early and late apoptotic or secondary necrotic) cells (Fig. 2b) (see Note 4).

5. The gating should be checked by analyzing keratinocytes incubated with 10 ng/ml CHX for 4 days, which shows three populations of live (quadrant 4), early apoptotic (quadrant 3), and necrotic cells (quadrant 2).
6. In quadrant 1, a fourth population of damaged viable cells could be detected depending on cell type and on centrifugation conditions.

4. Notes

1. Freshly thawed keratinocytes are very susceptible. A fast push out of the pipette-tip will kill them.
2. The incubation time for the trypsin solution can vary. It is important to check under the microscope that the cells are detached from the bottom before trypsin neutralization solution is added. Usually, it takes 3–5 min with freshly thawed trypsin solution. Trypsin solution should be stored in the freezer at −20°C and only aliquots should be thawed. For trypsinizing, trypsin solution should be heated up to 37°C to obtain the optimal enzyme activity. After thawing, it can be kept in the fridge up to 4 weeks and heated up to ten times without any noticeable activity loss.
3. It is recommended to use keratinocytes stimulated in 24- and 48-well plates to obtain an optimal cell number for flow cytometry analysis.
4. CHX-concentrations for apoptosis induction may vary in your set up. Do not be afraid to use up to 1 μg/ml CHX to induce cell death.
5. If keratinocytes are cultured in medium containing serum (e.g., transformed keratinocytes such as HaCaT), before trypsinizing a washing step should be performed since serum components inhibit trypsin.
6. After harvesting, centrifugation of keratinocytes should not be performed by less than 500 × *g* (dead cells could be lost).
7. Measure samples immediately by flow cytometry analysis and keep them in the dark at 4°C.

References

1. Kerr JF, Wyllie AH, Currie AR (1972). Apoptosis: a basic biological phenomenon with wide-ranging implications in tissue kinetics. Br J Cancer; 26:239–57.
2. Golstein P, Kroemer G (2007). Cell death by necrosis: towards a molecular definition. Trends Biochem Sci; 32:37–43.
3. Turcotte F, Lagasse P, Prud'homme D, Gendron R (1994). Regulated pre-employment medical examinations and their effect on the utilization of health services: the experience of 1,498 police cadets, Quebec, 1988. Can J Public Health; 85:348–50.
4. Abedin MJ, Wang D, McDonnell MA, Lehmann U, Kelekar A (2007). Autophagy delays apoptotic death in breast cancer cells following DNA damage. Cell Death Differ; 14:500–10.

Chapter 9

Measurement of Caspase Activity: From Cell Populations to Individual Cells

Gabriela Paroni and Claudio Brancolini

Abstract

Caspases are critical regulators of the apoptotic program, responsible for the harmonic dismantling of the cell. Cell death can occur by way of different options (necroptosis, necrosis, extreme autophagy) but once caspases are fully engaged it will take the apoptotic route. Hence, in general, caspase activity is inversely proportional to cell viability. Caspase activation can be measured by means of different techniques. Here, we describe three different methodologies to measure/observe caspase activation. Two approaches are recommended for studies on the whole cell population, whereas the third one is designed for visualizing caspase activation in single live cells.

Key words: Caspase, Apoptosis, Mitochondria, Bortezomib, Immunoblot, Time-lapse

1. Introduction

Caspases (cysteine proteases specific for Asp residues) are critical players in the cell death process. By cleaving a selected number of cellular proteins these enzymes are the master regulators of the type I programmed cell death (PCD) or apoptosis. Caspases are not only involved in the control of apoptosis. Additional biological functions, such as inflammation, innate immunity, and keratinocyte differentiation, are directed by nonapoptotic caspases. Moreover, it is now emerging that bona fide apoptotic caspases can also play specific roles outside of the one-way journey to peaceful death by apoptosis (1).

If caspases are unquestionably the master regulators of apoptosis (no-caspase implies no-apoptosis), cell death can still occur in the absence of these enzymes. Hence many death processes observed in the absence of detectable caspase activity or in the presence of caspase inhibitors have been commonly grouped as

Martin J. Stoddart (ed.), *Mammalian Cell Viability: Methods and Protocols*, Methods in Molecular Biology, vol. 740, DOI 10.1007/978-1-61779-108-6_9, © Springer Science+Business Media, LLC 2011

caspase-independent deaths. In many instances, but not all, the caspase-independent deaths take on the necrotic signature. Our knowledge on these caspase-independent deaths is scanty, mainly because of the few markers available to follow and discriminate the necrotic death process (2).

Caspases are synthesized as inactive zymogens and can be divided into initiators and effectors depending on the timing of activation. Initiator caspases (mainly 8 and 9) are monomers, activated earlier following adaptor-mediated oligomerization. Effector caspases (mainly 3, 6, and 7) are dimers, activated after cleavage mediated by initiator caspases. Once activated, effector caspases process selected death substrates and other caspases, in an amplificatory loop that causes cell fragmentation (1, 3).

Two main apoptotic pathways keep in check the caspase activation. The extrinsic pathway is triggered by the activation of a family of death receptors at the cell surface and controls the activation of caspase-8, whereas the intrinsic pathway is triggered by the release of killer mitochondrial proteins. Among these killers cytochrome *c* induces the assembly of the apoptosome and the activation of caspase-9.

Caspases are structurally organized into three domains: an N-terminal prodomain, a large subunit, and a small subunit. The mature enzyme results from processing of the zymogens between domains and the association of the two large and two small subunits to form the active tetramer (3). Caspase activation is considered a marker of apoptosis and can be measured by analyzing caspase processing via immunoblot or substrate cleavage via fluorogenic assays (see Subheadings 3.1 and 3.2). It is also possible to study the timing of caspase activation at the single cell level, in comparison with other apoptotic events, such as cell shrinkage, loss of plasma membrane integrity, and mitochondrial membrane potential decrease (see Subheading 3.3).

2. Materials

Before you make up any reagent or buffer, check the material safety data sheet (MSDS) that must be available in every laboratory for each chemical.

2.1. Cell Culture

1. Dulbecco's Modified Eagle's Medium (DMEM) (Lonza, Cambrex) supplemented with 10% fetal bovine serum (Euroclone), 2 mM glutamine (Lonza, Cambrex), and 100 U/ml penicillin–100 μg/ml streptomycin (Lonza, Cambrex). Store at 4°C.
2. Phosphate buffer saline (PBS). To prepare a 10× stock solution, dissolve 80 g NaCl, 2 g KCl, 14.4 g Na_2HPO_4, 2.4 g KH_2PO_4 in 800 ml distilled water; adjust to pH 7.4 and to a

total volume of 1 l; sterilize by autoclaving. Store at room temperature (RT).

3. Trypsin–EDTA.
4. Bortezomib (1 μg/ml) (LC Laboratories).
5. Etoposide (50 μg/ml) (Sigma).
6. Cell culture-treated flasks and Petri dishes.

2.2. Cell Lysis

1. PBS.
2. 1 M Tris–HCl, pH 6.8. Dissolve 12.1 g of Tris base in 80 ml of distilled water. Adjust the pH to 6.8 with concentrated HCl. Adjust the volume to 100 ml. Store at RT.
3. 10% SDS. Dissolve 10 g of sodium-dodecyl-sulfate (SDS) in 1 l of distilled water.
4. Laemmli sample buffer. To prepare a 2× stock solution, combine 4 ml of 10% SDS, 2 ml of glycerol, 1.2 ml of 1 M Tris–HCl, pH 6.8, and 0.01% bromophenol blue. Finally, adjust to 10 ml with distilled water. Supplement the buffer with 2% β-mercaptoethanol prior to use. Store at RT.
5. Teflon cell scrapers.

2.3. SDS–PAGE

1. 30% Acrylamide/bis solution (29:1). Store at 4°C.
2. 1.5 M Tris–HCl, pH 8.8. Dissolve 181.5 g of Tris base in 80 ml of distilled water. Adjust the pH to 8.8 with concentrated HCl. Adjust the volume to 1 L. Store at RT.
3. 1 M Tris–HCl, pH 6.9. Dissolve 12.1 g of Tris base in 80 ml of distilled water. Adjust the pH to 6.9 with concentrated HCl. Adjust the volume to 100 ml. Store at RT.
4. *N*,*N*,*N*′,*N*′-Tetramethylethylenediamine (TEMED).
5. Ammonium persulfate. Prepare a 10% solution in water and freeze it immediately in aliquots ready for use. Store at –20°C.
6. Running buffer. To prepare a 10× stock solution, dissolve the following compounds in 1 l of distilled water: 30 g Tris base, 144 g glycine, 10 g SDS. Store at RT.
7. Molecular weight markers (14 kDa up to 200 kDa, Sigma).

2.4. Immunoblotting

1. Semidry transfer buffer. Add 5.82 g Tris base, 2.93 g glycine, 3.75 ml of 10% SDS to 800 ml of distilled water. Adjust to 1 l with 200 ml of methanol. Store at RT.
2. Nitrocellulose membrane and 3MM chromatography paper.
3. Ponceau solution: Dissolve 0.1% (w/v) Ponceau red in 5% acetic acid solution. Store at RT.
4. 5 M NaCl solution. 292.2 g of NaCl in 800 ml of distilled water. Adjust the volume to 1 l with distilled water. Sterilize by autoclaving. Store at RT.

5. 1 M Tris–HCl, pH 7.5. Dissolve 121.1 g of Tris base in 800 ml of distilled water. Adjust the pH to the desired value by adding concentrated HCl.
6. Blocking buffer. Combine 20 ml 5 M NaCl, 25 ml 1 M Tris–HCl, pH 7.5, and 0.5 ml Tween-20 in 500 ml of distilled water and dissolve no-fat dry milk to a final concentration of 5% (w/v). Store at 4°C.
7. Primary antibodies: anti-caspase-3 and caspase-7 (Signal Transduction), anti-caspase-8 (Alexis), anti-PARP (Promega), anti-caspase-2 (4), anti-caspase-9 (5).
8. Secondary antibodies: anti-mouse or anti-rabbit peroxidase-conjugated.
9. Detection system: Super signal west pico chemiluminescent substrate (Thermo Scientific).
10. Autoradiography films (Amersham hyperfilm GE Healthcare) or alternatively the molecular imager Chemidoc (Bio-Rad).

2.5. Fluorogenic Caspase Assay

1. Fluorescent plate reader (Molecular Device).
2. Multichannel pipettors.
3. Plate shaker.
4. Apo-ONE® Caspase-3/7 reagent (Promega).

2.6. Time-Lapse Analysis

1. 35 mm Glass bottom Petri dishes (Wilco Wells B.V.).
2. GFP-DD-TM, GFP-AA-TM expressing plasmids (6).
3. Transfection reagent: Lipo2000 (Invitrogen) or alternatively and automated microinjection system (Zeiss).
4. Cell culture tested mineral oil (Sigma).
5. Confocal laser scanning microscope (Leica TCS-SP).
6. Time-lapse apparatus: Incubators for microscopes S-2 or ML-Leica (Pecon), device supplying a heated CO_2–air mixture and a temperature controller (CTI-Controller 3700 digital and Tempcontrol 37-2 – Pecon).
7. Software for imaging analysis (MetaMorph).

3. Methods

Caspase activation can be analyzed by immunoblot using antibodies able to recognize both the pro-form and the active enzyme (processed domains) or the caspase-cleaved substrates. A list of caspase substrates (The Casbah) is available at http://www.casbah.ie database (7).

Since the apoptotic process is an asynchronous event, it is advisable to collect samples at different time points from the addition of the apoptotic trigger. Generally, the ratio between the pro-form and the cleaved-form of a caspase or of a caspase-substrate can be considered a parameter of the apoptotic magnitude (Fig. 1). However, the simultaneous measurement of cell death using alternative methods such as subG1 content (with a cytofluorimeter) or the trypan-blue assay, which scores plasma membrane integrity, is helpful to corroborate the rate of caspase or of the caspase-substrate processing.

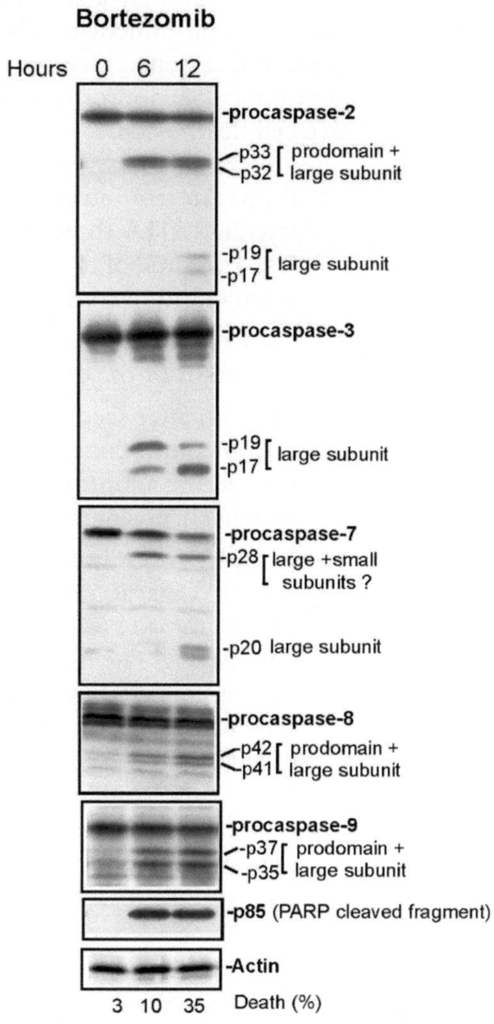

Fig. 1. Evaluating caspase activation by immunoblot. Human fibroblasts expressing the oncogenes E1A, Ras, and a dominant negative mutant of p53 were treated for the indicated times with the proteasome inhibitor bortezomib (1 μM). Processing of the indicated caspases was analyzed by using the following antibodies and dilutions: anti-caspase-2 (1/2,000) (4), anti-caspase-3 (1/1,000), anti-caspase-7 (1/1,000), anti-caspase-8 (1/2,000), anti-caspase-9 (1/1,000) (5), anti-cleaved PARP (1/2,000). Equal amounts of cell lysates were subjected to SDS/PAGE electrophoresis. The appearance of cell death was scored by trypan blue staining.

It is important to take into account that apoptotic cells undergo clearance from neighboring cells or swell and release components, in a process known as secondary necrosis. These events and the asynchronous trait of the apoptotic process could impair the detection of the caspase-cleaved forms, in a cell population. In the case of adherent cells, collecting viable adherent cells and floating dead cells separately can overcome this problem. Moreover, it should be kept in mind that the caspase-cleaved forms are inherently labile. Hence, when samples are collected at a time point late after the induction of apoptosis, secondary necrosis could play a relevant part thus affecting the majority of the cells and influencing the detection of the caspase-processed peptide-fragments. Sample collection at later times (after the appearance of apoptosis) should be avoided, especially when massive cell death occurs.

3.1. Measurement of Caspase Activation by Immunoblot

3.1.1. Sample Preparation for Scoring Caspase Activation by Immunoblot

1. Propagate the adherent cell culture by splitting cells with trypsin–EDTA three times a week or as specifically required. Use a volume of trypsin–EDTA just enough to cover the cells when culture vessel is tilted.
2. Seed $1–2 \times 10^5$ cells/ml in a 60-mm Petri dish for each single data point of the experiment (see Note 1).
3. 24–48 h after seeding change the medium and add fresh culture medium containing the vehicle alone (as a control) or the specific apoptotic trigger. Make sure that drugs are given to actively proliferating cells.
4. In order to evaluate the appearance of apoptosis over time, harvest cells at different time points from drug addition. Check for the effective appearance of apoptosis by light microscopy. The media containing the floating apoptotic cells can be collected in a 15-ml falcon tube. Wash the adherent cells with 5 ml of PBS. After washing add the 5 ml of PBS (which should contain poorly attached cells and debris) to the falcon tube containing the floating apoptotic cells. Adherent cells can now be scraped into 1 ml of ice-cold PBS. At this point, it is possible to decide whether to keep the two samples separate or, the healthy adherent cells can be mixed with the floating dead cells. When apoptosis is induced at low levels it is recommended to keep the two samples separated. Finally, since caspase processed forms are inherently labile, samples should be harvested rapidly and kept on ice.
5. Centrifuge the cell suspension 10 min at $800 \times g$, discard the supernatant, and wash the resulting pellet with ice-cold PBS, to remove serum. Resuspend the pellet in PBS and transfer it to an Eppendorf tube. Centrifuge again (10 min at $800 \times g$) and resuspend the pellet in Laemmli sample buffer (1×). For the untreated cells, that should reach the confluence at the harvesting time, add 300 μl of Laemmli sample buffer.

An adequate volume of Laemmli sample buffer should be added to the treated samples proportionally to the pellet volume or the cell number.

6. DNA can be sheared through sonication (Bioruptor-Diagenode) or by the use of a 29-gauge insulin-needle syringe. If you are dealing with many samples sonication is preferred for its reproducibility.
7. Samples are boiled for 5 min and after cooling to room temperature they are ready for separation on SDS-PAGE electrophoresis. Keep lysates in ice during loading procedure (minimize the time) and at –80°C for long-term storage (see Note 2).

3.1.2. SDS-PAGE

We describe the procedure for running a SDS-PAGE electrophoresis using the Mini-Protean 3 apparatus (Bio-Rad). Volumes can change accordingly for different suppliers.

1. Assemble the glass and spacers for the polymerization of a 0.75-mm thick minigel.
2. Prepare 5 ml of a 15% running gel by combining 1.1 ml distilled water, 2.5 ml 30% acrylamide-bis solution (29:1), 1.3 ml 1.5 M Tris–HCl solution pH 8.8, and 50 μl 10% SDS. Induce gel polymerization by adding 50 μl of 10% ammonium persulfate solution and 2 μl TEMED. Pour the gel, leaving space for the stacking gel (1:5 of the glass height), and gently overlay with water. The running gel should polymerize in about 20 min.
3. Pour off the water and prepare 2 ml of stacking gel by combining 1.4 ml distilled water, 0.33 ml 30% acrylamide-bis solution (29:1), 0.25 ml 1 M Tris–HCl solution, pH 6.8, and 20 μl 10% SDS. Induce gel polymerization by adding 20 μl of 10% ammonium persulfate solution and 2 μl TEMED. Pour the stacking gel and gently insert the comb avoiding bubbles. The gel should polymerize in about 20 min.
4. Prepare the running buffer by diluting the 10× stock running buffer in a measuring cylinder. Cover with parafilm and invert to mix.
5. Once the gel has completely polymerized, carefully remove the comb, set the gel in the electrophoresis apparatus, and fill the chambers with the running buffer 1×. Wash the wells with a 2.5-ml syringe fitted with a 21-gauge needle.
6. Load the samples and the molecular weight markers.
7. Complete the assembly of the electrophoresis unit and connect it to a power supply. Run the gel at 80 V during the stacking and 180 V during the running gel. The run should last one and a half hours.

3.1.3. Measurement of Caspase Activation by Immunoblot

Proteins from the gel can be transferred to the nitrocellulose membrane both by wet and semidry electrophoretic transfer procedures. Here, we describe the semidry transfer method, which is quicker and useful for proteins with molecular weight comparable to caspases and their cleaved fragments. For high molecular weight proteins, such as particular caspase substrates, the wet apparatus is highly recommended.

1. Prepare six 3MM paper sheets of the same dimension of the running gel (approximately 5.5 × 8.5 cm), soak them in the semidry transfer buffer.
2. Assemble the blotting sandwich on the bottom (anode) plate of the apparatus, starting by laying three 3MM papers, next roll over the sheets with a Pasteur pipette to eliminate excess buffer.
3. Disassemble the gel apparatus and remove the stacking gel. Wash the resolving gel in water and submerge it in a container of transfer buffer.
4. Soak carefully the nitrocellulose membrane (5.5 × 8.5 cm) in transfer buffer (it wets by capillary action) and lay it on the top of three 3MM sheets.
5. Carefully put the gel on the top of the nitrocellulose membrane.
6. Place the three remnant 3MM sheets on the top of the gel, ensuring that no bubbles are trapped in the resulting sandwich. To eliminate excess buffer and bubbles roll over the gel-paper sandwich with a Pasteur pipette.
7. Close the semidry apparatus and allow the transfer to occur according to the apparatus manual instruction (usually 60 min with a current of 0.8 mA/cm^2 is suggested). Remember: proteins will migrate from the cathode (−) to the anode (+) verify that the sandwich has been assembled correctly.
8. To verify that the transfer has properly occurred, disassemble the sandwich, wash the membrane in water, and next soak it in the Ponceau solution for 1 min. Wash the membrane in water again. If the transfer had occurred correctly, proteins will be red stained uniformly, in the different lanes (see Note 3).
9. The membrane is then incubated in the blocking buffer for 1 h at room temperature on a rocking platform.
10. Next incubate the membrane in blocking buffer supplemented with the desired primary antibody for 1 h at room temperature or overnight at 4°C.
11. The primary antibody is then removed (see Note 4) and the membrane is washed three times with the blocking buffer for 10 min.

12. Incubate the membrane in blocking buffer supplemented with the secondary antibody (horseradish peroxidase-conjugated) for 1 h at room temperature.
13. The secondary antibody solution is then discarded and the membrane is washed three times for 10 min in blocking buffer (see Note 5).
14. After a final wash in PBS, the membrane is ready for the chemiluminescence detection.
15. The SuperSignal reagents are mixed just prior to use according to the manufacturer's instructions. Prepare a volume enough to wet and cover the membrane (usually 2 ml per membrane) and place the liquid onto parafilm placed on a smooth surface (the bench). The membrane is laid on the solution, paying attention that the protein side of the membrane is completely covered by the SuperSignal solution.
16. After a few seconds, remove the blot from the SuperSignal solution, absorb the excess liquid with a 3MM paper, and place it within an acetate sheet protector.
17. While in a darkroom, place the acetate protector containing the membrane in an X-ray film cassette and cover the membrane with an autoradiography film for a suitable exposure time, typically from few seconds to some minutes. An example of an immunoblot procedure with anti-caspase antibodies is shown in Fig. 1. Alternatively, the chemiluminescent signal can be captured digitally with a high quality detector provided with a CCD camera, such as the molecular imager Chemidoc (Bio-Rad) (see Note 6).

3.2. Measurement of Caspase Activation Using Fluorimetric Assays

Measurement of active caspase by using fluorescent probes can be an alternative strategy to evaluate the induction of apoptosis in a cell population. This assay can be chosen to corroborate the immunoblot data on caspase activation. Moreover, when it is coupled to a trypan blue analysis, which simply score cell death without any hints on the kind of death process, it proves that death is indeed apoptotic, with a very fast and simple procedure. Measurement of active caspases is also an ideal assay for high-throughput screening.

The best assays are based on probes that become fluorescent after caspase cleavage. They are highly sensitive so few cells are required to get a good signal, hence the assay can be performed in 96-, 375-well plates culture. These probes usually contain a tetra(penta) peptide mirroring the consensus site for caspase processing (DXXD for caspase-3 and 7) that allows the release of a fluorophore after caspase activation (8). Many of these reagents are commercially available as kits or as fluorescent peptides. In the laboratory, we routinely use the Apo-ONE® homogeneous caspase-3/7 assay (Fig. 2).

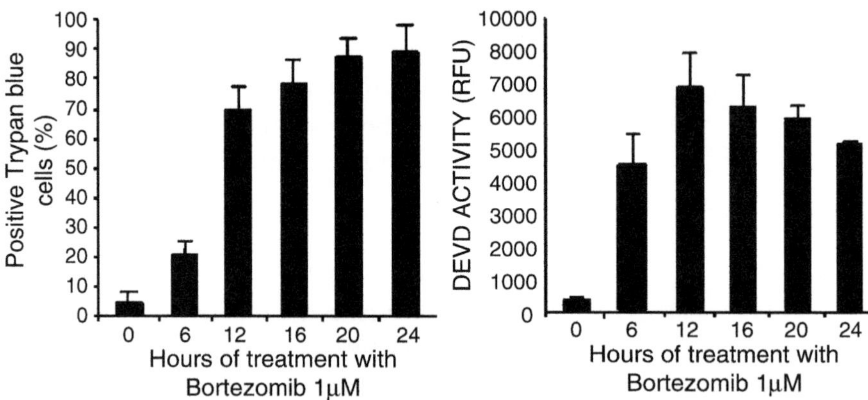

Fig. 2. Evaluating caspase activation by scoring caspase-3/-7 (DEVDase) enzymatic activity. (**a**) Human fibroblasts expressing the E1A oncoprotein were treated for the indicated times with 1 μM of the proteasome inhibitor bortezomib. The appearance of apoptosis was scored by trypan blue staining (mean ± SD *n* = 3). (**b**) Human fibroblasts expressing the E1A oncoprotein were treated for the indicated times with 1 μM of bortezomib and caspase-3/-7 activities have been measured with the Apo-ONE® kit. Some decrease in DEVDase activity can be appreciated at later time point from apoptosis induction (mean ± SD *n* = 3).

This assay includes a profluorescent caspase-3/7 consensus substrate: rhodamine 110, bis-(*N*-CBZ-L-aspartyl-L-glutamyl-L-valyl-aspartic acid amide; Z-DEVD-R110). Upon cleavage on the C-terminal side of the aspartate residue and removal of the DEVD peptide operated by caspase-3/7 enzymes, the rhodamine 110 group becomes intensively fluorescent.

The assay includes a cell lysis buffer, which supports caspase activity. Hence, the reagent is added directly to samples. It should be kept in mind that since the amount of fluorescent product generated is representative of the amount of active caspase-3/7 present in the sample, scoring apoptosis at later times from its induction will result in a decrease in caspase activity, and thus in fluorescent signal, due to apoptotic cell clearance or secondary necrosis.

1. For the experiment in 96-well plates, at least a triplicate is necessary for each data point.
2. Seed 100 μl of cell suspension at a concentration of 1×10^5 cell/ml 24 h before drug addition.
3. Prepare the following data points:
   (a) Blank: Cell culture medium without cells.
   (b) Negative control: Vehicle-treated cells.
   (c) Samples: Treated cells.

   The blank control is used as a measure of background fluorescence associated with the cell culture medium and the Apo-ONE® Caspase-3/7 reagent. It should be subtracted from the experimental values. Negative control reactions are useful for determining the basal caspase activity of the cell line.

In order to obtain a kinetic evaluation of the process, it is suggested to analyze caspase activity at different time frames from the apoptotic insult.

4. Add 100 μl of Apo-ONE Caspase-3/7 reagent to each well of the 96-well plate containing 100 μl of media.
5. Gently mix the contents of the wells by using a plate shaker at 300–500 rpm for 30 s.
6. Incubate at room temperature in the dark for 30 min to 18 h depending upon expected level of apoptosis.

   The optimal incubation period should be determined empirically. Minimal apoptotic induction and low cell number may require an extended incubation time. Maximum recommended incubation time is 18 h. Cover the plate with a plate sealer if incubating for extended periods is required.
7. Repeatedly measure the fluorescence of each well over time.

   The optimal emission and excitation wavelength depend on the fluorophore. As explained above, the fluorophore in the Apo-ONE® kit is rhodamine 110, which can be detected with an excitation wavelength range of 485 ± 20 nm and an emission wavelength range of 530 ± 25 nm. The amount of fluorescent product generated is proportional to the amount of caspase-3/7 cleavage activity present in the sample (see Note 6).

3.3. Measurement of Caspase Activation at Single Cell Level by Time-Lapse Analysis

Caspase activation can be observed and measured also at the single cell level. Investigating caspase activation at the single cell level is important to understand the reciprocal dependences among cellular changes characterizing apoptosis and the timing of activation of these enzymes. Several groups have developed probes and methods to monitor caspase activation in single cells. Generally, these probes make use of the fluorescent resonance energy transfer (FRET) technology. Cyan fluorescent protein (CFP) and yellow fluorescent protein (YFP) are linked by a short peptide containing different caspases consensus cleavage sites. A decreased FRET activity results from the cleavage of the linker peptide and it can be monitored as an indication of caspase activity (9).

We have used a different and simplified approach by generating a GFP fusion protein (TM-DD-GFP) harboring a mitochondrial transmembrane domain (TM) and a consensus sequence for caspase cleavage (DD). This fusion represents a specific caspase-3/-7 sensor located close to the core of the apoptotic machinery: the outer mitochondrial membrane. Once caspases are activated, the GFP domain is released from the outer mitochondrial membrane to the cytosol (5). Hence, there is a drastic redistribution of the GFP fluorescence from a punctuate (mitochondria) to a diffuse staining (cytoplasm). This redistribution of GFP fluorescence is an indirect marker of caspase activity.

As a paradigm of studying the apoptotic process at a single cell level, we report the simultaneous analysis of mitochondrial membrane potential ($\Delta\psi_m$) dissipation and of caspase activation, after treatment of IMR90-E1A cells with the DNA damaging agent etoposide (Fig. 3).

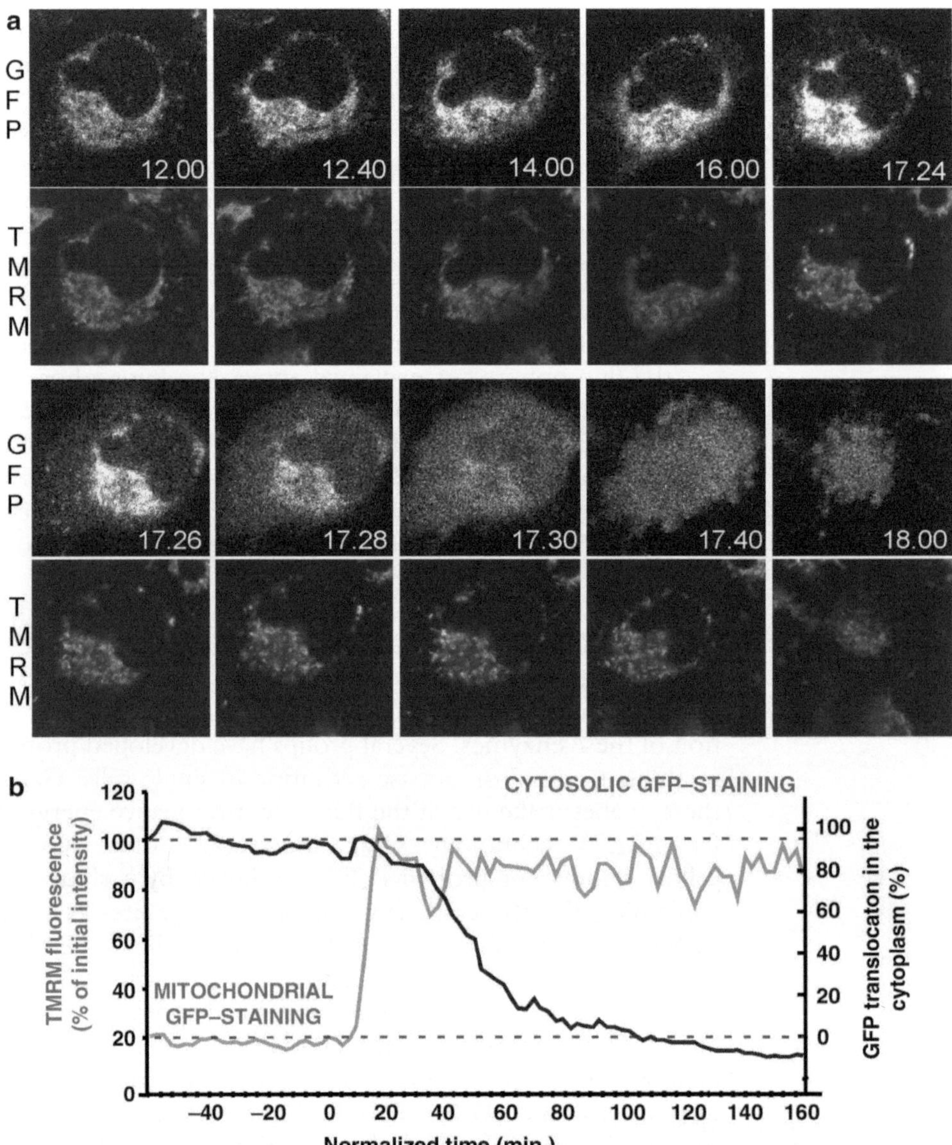

Fig. 3. Evaluating caspase activation in vivo at single cell level. (**a**) Selected time-lapse images of human fibroblasts transformed with E1A oncogene (E1A cells) and expressing the caspase-3 probe GFP-DD-TM. Cells were incubated with etoposide (50 μM) and TMRM (20 nM) to evaluate mitochondrial membrane potential. Times indicate hours after etoposide addition. Note the translocation of GFP fluorescence from a punctuate staining (mitochondria) to a diffuse cytosolic activation which occurs upon caspase-3 activation. (**b**) Cytosolic translocation of GFP-DD-TM and decrease of mitochondrial TMRM uptake in single cells as markers of caspase activation and $\Delta\psi_m$ dissipation. Individual trace for TMRM uptake and GFP localization of a typical E1A cell expressing GFP-DD-TM after treatment with 50 μM etoposide.

3.3.1. Expression of the Caspase-3/-7 Reporter in Mammalian Cells

1. Generate cell lines expressing the GFP-probe. Transient transfection microinjection, or the isolation of stable transfected cell lines are all useful strategies to express the GFP-based caspase-3/-7 sensors in the cell line of interest.
2. Seed cells from 24 to 48 h before transfection or microinjection according to the cell type and the optimal transfection or microinjection conditions. For this experiment, 35-mm glass bottom dishes are required.
3. Microinject or transfect the cells with the TM-DD-GFP expressing plasmid. For microinjection a solution of 2 ng/μl is required. For transient transfection an amount of plasmid varying from 100 to 300 ng/35 mm dish is recommended. As a control, transfect cells with a plasmid expressing the same fusion protein mutated in the consensus caspase cleavage site (TM-AA-GFP).
4. 3 h after microinjection or approximately 12–18 h from transfection, pre-incubate the cells for 2 h with the $\Delta\psi_m$ sensitive dye tetramethyl rhodamine methyl-ester (TMRM) and next treat them with the apoptotic inducer. Mitochondrial depolarization associates with the decrease of TMRM uptake. The optimal TMRM concentration should be determined empirically for each cell line used in your studies.

3.3.2. Time Lapse

1. Place the dish into the confocal microscope equipped with a time-lapse apparatus. If the chamber does not include a humidifier the media must be covered with cell-tested mineral oil to prevent evaporation.
2. Acquire images using a 63× oil fluorescence objective every 2 min. To detect the GFP fluorescent fusion protein and TMRM, use the excitation wavelength of a 488 nm Argon laser and 543 nm HeNe laser, respectively.

3.3.3. Measure the $\Delta\psi_m$ Decrease by Single Cell Image Analysis

1. Use MetaMorph or similar software to perform image analysis.
2. For TMRM analysis in single cells draw a region (ROI) around fluorescent mitochondria (MT) and measure the total brightness (integrated fluorescence intensity or II) of such a region (II_{TMRM}).
3. Subtract the background from the II_{TMRM} (Bk_{TMRM}), acquired by using a ROI positioned in a nonfluorescent region, according to the following equation:

$$BK_{TMRM} = (II_{ROI} / Area_{ROI}) \times Area_{MT}$$

4. Represent the background-corrected II_{TMRM} values as the percentage of the average of at least 29 values detected before the II_{TMRM} started decreasing. The II_{TMRM} values have to be

taken for 60 min before and 120/240 min after the beginning of the $\Delta\psi_m$ decrease (set to min 0).

3.3.4. Measure Caspase Activation by Single Cell Image Analysis

5. Calculate the kinetics of GFP release by measuring the standard deviation (SD) from the average pixel intensity of a ROI positioned on the mitochondria and cytosol of individual cells (II_{GFP}).
6. Subtract the background from the II_{GFP} (Bk_{GFP}) image by using a ROI positioned in a nonfluorescent region according to the following equation:

$$BK_{GFP} = (II_{ROI} / Area_{ROI}) \times Area_{GFP}$$

   Mitochondrial localized TM-DD-GFP (punctuate) contributes to a high SD of pixel intensities, whereas cytosolic, homogeneously distributed, GFP (diffuse) is represented by a low SD.
7. To better examine the relationship between $\Delta\psi_m$ loss and caspase activation, illustrate the punctuate–diffuse GFP values as the inverse value of SD after background subtraction ($1/SD = y$). For direct comparisons and statistical analyses, transform the single cell standard deviation traces in a 0–100% scale in which 0% equals to the first y punctuate value (y_{MT}) and 100% equals the first complete diffusion value (y_{diff}) according to the following equation:

$$y_{(0-100\%)} = 100 \times \left(\frac{y - y_{MT}}{y_{diff} - y_{MT}} \right)$$

   A representation of the data obtained with this analysis is shown in Fig. 3.

4. Notes

1. The number of cells to be seeded for each experiment should be evaluated according to the growth rate of the cell line used. Pro-apoptotic drugs should be administrated to proliferating cells and sample collection should occur before untreated cells reach over-confluence.
2. To avoid repeated freeze and thaw cycle of the extracts, prepare some aliquots of the cellular lysates before storing them at –80°C.
3. A superior immunoblot depends on the quality of the antibody and the concentration of the cellular lysates. After the transfer to the membrane, the Ponceau staining can offer an important test for the quality of the immunoblot. Lanes containing the cellular proteins should be uniformly and intensely red-colored respect to a white background.

4. The primary antibody can be saved for subsequent experiments by adding sodium azide (0.02% final concentration) and storing at 4°C (caution azide is highly toxic).
5. It is advisable to try a titration experiment to determine the optimal antibody dilution (for both primary and secondary antibodies), which can change for cell line to cell line and from the amount of proteins in the cellular lysates.
6. A broad caspase inhibitor such as Z-VAD.fmk (Z-VAD-fluoromethylketone, Bachem) can be used to definitively prove that the proteolytic cleavage of a death substrate or even of a caspase, detected in immunoblots, is triggered by a caspase activity. Z-VAD.fmk can also be employed as a control for the fluorogenic assay.

Acknowledgments

This work is supported by AIRC (Associazione Italiana Ricerca sul Cancro), MIUR (Ministero dell'Istruzione, Università e Ricerca), and Regione Friuli-Venezia Giulia (AITT-LR25/07).

References

1. Caroline, H. Y., and Yuan, J. (2009) The Jekyll and Hyde Functions of Caspases *Dev. Cell* **16**, 21–34
2. Zong W.X., and Thompson, C.B. (2006) Necrotic death as a cell fate. *Genes Dev.* **20**, 1–15
3. Pop, C., and Salvesen, G.S. (2009) Human caspases: activation, specificity, and regulation. *J Biol Chem.* **284**, 21777–81
4. Paroni, G., Henderson, C., Schneider, C., and Brancolini, C. (2001) Caspase-2-induced apoptosis is dependent on caspase-9, but its processing during UV- or tumor necrosis factor-dependent cell death requires caspase-3. *J Biol Chem.* **276**, 21907–1
5. Rodriguez J, Lazebnik Y. (1999) Caspase-9 and APAF-1 form an active holoenzyme. *Genes Dev.* **13**, 3179–84
6. Henderson, C.J., Aleo, E., Fontanini, A., Maestro, R., Paroni, G., and Brancolini C. (2005) Caspase activation and apoptosis in response to proteasome inhibitors. *Cell Death Differ.* **12**, 1240–54
7. Lüthi, A.U., and Martin, S.J. (2007) The CASBAH: a searchable database of caspase substrates. *Cell Death Differ.* **14**, 641–50
8. Garcia-Calvo, M., Peterson, E.P., Leiting, B., Ruel, R., Nicholson, D.W., and Thornberry, N.A. (1998) Inhibition of human caspases by peptide-based and macromolecular inhibitors. *J Biol Chem.* **273**, 32608–13
9. Rehm, M., Dussmann, H., Janicke, R.U., Tavare, J.M., Kogel, D., Prehn, J.H. (2002). Single-cell fluorescence resonance energy transfer analysis demonstrates that caspase activation during apoptosis is a rapid process. Role of caspase-3. *J Biol Chem.* **277**, 24506–14

Chapter 10

Rapid Quantification of Cell Viability and Apoptosis in B-Cell Lymphoma Cultures Using Cyanine SYTO Probes

Donald Wlodkowic, Joanna Skommer, and Zbigniew Darzynkiewicz

Abstract

The gross majority of classical apoptotic hallmarks can be rapidly examined by multiparameter flow cytometry. As a result, cytometry became a technology of choice in diverse studies of cellular demise. In this context, a novel class of substituted unsymmetrical cyanine SYTO probes has recently become commercially available. Derived from thiazole orange, SYTO display low intrinsic fluorescence, with strong enhancement upon binding to DNA and/or RNA. Broad selection of excitation/emission spectra has recently driven implementation of SYTO dyes in polychromatic protocols with the detection of apoptosis being one of the most prominent applications In this chapter, we outline a handful of commonly used protocols for the assessment of apoptotic events using selected SYTO probes (SYTO 16, 62, 80) in conjunction with common plasma membrane permeability markers (PI, YO-PRO 1, 7-AAD).

Key words: Flow cytometry, Apoptosis, Lymphoma, SYTO 16, SYTO 62, SYTO 80

1. Introduction

During the past decade, mechanisms underlying cell death have entered into a focus of many researchers in diverse fields of biomedicine. The gross majority of classical apoptotic hallmarks can be rapidly examined by multiparameter flow cytometry (1). As a result, cytometry became a technology of choice in diverse studies of cellular demise, and diverse cytometric assays have been introduced (1–3). To date, however, live-cell assays that are based on cell permeant DNA probes suffered mostly from their unfavorable spectral characteristics that necessitate UV excitation source and dedicated optics. Excessive toxicity/phototoxicity precluded also long-term studies such as cell sorting with subsequent cell cultivation (4, 5).

Martin J. Stoddart (ed.), *Mammalian Cell Viability: Methods and Protocols*, Methods in Molecular Biology, vol. 740, DOI 10.1007/978-1-61779-108-6_10,

In this context, a novel class of substituted unsymmetrical cyanine SYTO probes has recently become commercially available. Derived from thiazole orange, SYTO display low intrinsic fluorescence, with strong enhancement upon binding to DNA and/or RNA. This novel class of probes spans a broad range of visible excitation and emission spectra: (1) SYTO blue (Ex/Em 419–452/445–484 nm); (2) SYTO green (Ex/Em 483–521/500–556 nm); (3) SYTO orange (Ex/Em 528–567/544–583 nm); and (4) SYTO red (Ex/Em 598–654/620–680 nm) (4, 6–8). Exploitation of SYTO probes to cytometric detection of apoptosis is a relatively new method (4, 6–8). This methodology, however, is slowly gaining appreciation as an easy to perform, live-cell assay (6–8). Importantly, we have recently showed that cyanine SYTO dyes represent a promising class of inert probes that do not adversely affect normal cellular physiology (9). When preloaded or continuously present in medium, SYTO do not interfere with cell viability, and their intracellular retention permits straightforward and kinetic analysis of investigational drug/compound cytotoxicity (9). Reduction of sample processing achieved with these protocols is important for preservation of fragile apoptotic cells. SYTO 16-based live cell sorting represents also a novel approach to supravitally track progression of apoptotic cascade in response to investigational anticancer agents (9).

In this chapter, we outline a handful of commonly used protocols for the assessment of apoptotic events using selected SYTO probes (SYTO 16, 62, 80) in conjunction with common plasma membrane permeability markers (PI, YO-PRO 1, 7-AAD).

2. Materials

2.1. Detection of Apoptosis Using SYTO 16 and Propidium Iodide

1. Cell suspension (2.5×10^5–2×10^6 cells/mL).
2. 1× PBS.
3. 1 mM SYTO 16 stock solution in DMSO. Store protected from light at −20°C. Stable for over 12 months. Caution: although there are no reports on SYTO 16 toxicity, appropriate precautions should always be applied when handling SYTO 16 solutions.
4. 10 μM SYTO 16 working solution in PBS (prepare fresh as required).
5. 50 μg/mL propidium iodide (PI) stock solution in PBS. Store protected from light at +4°C. Stable for over 12 months. Caution: PI is a DNA binding molecule and thus can be considered as a potential carcinogen. Always handle with care and use protective gloves.

6. 30 mM Verapamil in EtOH (P-gp inhibitor). Store protected from light at −20°C. Stable for over 12 months.
7. 1.5 mL Eppendorf tubes.
8. 12 × 75 mm polystyrene FACS tubes.

2.2. Detection of Apoptosis Using SYTO 62 and YO-PRO 1

1. Cell suspension (2.5×10^5–2×10^6 cells/mL).
2. 1× PBS.
3. 1 mM SYTO 62 stock solution in DMSO. Store protected from light at −20°C. Stable for over 12 months. Caution: although there are no reports on SYTO 62 toxicity, appropriate precautions should always be applied when handling SYTO 62 solutions.
4. 10 μM SYTO 62 working solution in PBS (prepare fresh as required).
5. 1 mM YO-PRO 1 stock solution in DMSO. Store protected from light at −20°C. Stable for over 12 months. Caution: although there are no reports on YO-PRO 1 toxicity, appropriate precautions should always be applied when handling YO-PRO 1 solutions.
6. 10 μM YO-PRO 1 working solution in PBS (prepare fresh as required).
7. 30 mM Verapamil in EtOH (P-gp inhibitor). Store protected from light at −20°C. Stable for over 12 months.
8. 1.5 mL Eppendorf tubes.
9. 12 × 75 mm polystyrene FACS tubes.

2.3. Detection of Apoptosis Using SYTO 80 and 7-AAD

1. Cell suspension (2.5×10^5–2×10^6 cells/mL).
2. 1× PBS.
3. 1 mM SYTO 80 stock solution in DMSO. Store protected from light at −20°C. Stable for over 12 months. Caution: although there are no reports on SYTO 80 toxicity, appropriate precautions should always be applied when handling SYTO 80 solutions.
4. 10 μM SYTO 80 working solution in PBS (prepare fresh as required).
5. 50 μg/mL 7-aminoactinomycin D (7-AAD) stock solution in PBS. Store protected from light at +4°C. Stable for over 12 months. Caution: 7-AAD is a DNA binding molecule and thus can be considered as a potential carcinogen. Always handle with care and use protective gloves.
6. 30 mM Verapamil in EtOH (P-gp inhibitor). Store protected from light at −20°C. Stable for over 12 months.
7. 1.5 mL Eppendorf tubes.
8. 12 × 75 mm polystyrene FACS tubes.

3. Methods

3.1. Detection of Apoptosis Using SYTO 16 and PI

The cytometric detection of SYTO 16 fluorescence loss is a sensitive marker of early apoptotic events (Fig. 1a; see Notes 1–3) (4, 7, 9). The following protocol describes a combined use of SYTO 16 (Ex/Em 488/518 nm) together with a marker of plasma membrane integrity: propidium iodide (PI; Ex/Em 488/575–670 nm). Every assay allows for sensitive discrimination between viable, early apoptotic, and late apoptotic/necrotic subpopulations based on differential SYTO staining profiles (Fig. 1a; see Notes 3–5) (4, 9). The method presented below is a single-step and time-saving assay. Minimized washing steps permit maintenance of fragile apoptosing population in intact state (see Note 5) (4, 7).

1. Collect cell suspension into 12 × 75-mm Falcon FACS tube and centrifuge for 5 min, 243 × *g* at room temperature (RT).
2. Prepare staining mixture by adding 875 μL of PBS, 25 μL of 10 μM SYTO 16, and 100 μL of 50 μg/mL PI (final concentration 250 nM SYTO 16 and 5 μg/mL PI).
3. Add 1 μL of 30 mM Verapamil to the staining mixture (final concentration 30 μM; see Note 6).
4. Discard supernatant and gently resuspend cells in 100 μL of staining mixture.
5. Incubate for 15 min at RT.
6. Add 500 μL of PBS.
7. Analyze on flow cytometer with 488 nm excitation line (Argon-ion laser or solid-state laser) with emissions collected at 530 nm (SYTO 16) and 575–610 nm (PI). Adjust the logarithmic amplification scale to distinguish between viable cells (bright SYTO $16^+/PI^-$), early apoptotic cells (dim SYTO $16^+/PI^-$), and late apoptotic and/or necrotic cells with compromised plasma membranes (low SYTO $16^+/PI^+$) (Fig. 1a; see Notes 7–9).

3.2. Detection of Apoptosis Using SYTO 62 and YO-PRO 1

The principle of this assay is similar to the above described SYTO 16 protocol (see Notes 1–3). The main advantage of this protocol is the minimal interbeam compensation required between SYTO 62 and YO-PRO 1 (see Note 10) (4, 8). Moreover, the use of YO-PRO 1 (Ex/Em 491/509 nm) as a marker of plasma membrane integrity allows for a more sensitive detection of early apoptotic cells than propidium iodide (8, 10). Furthermore, other channels from 488 nm excitation line are free to combine this protocol with, e.g., immunophenotyping.

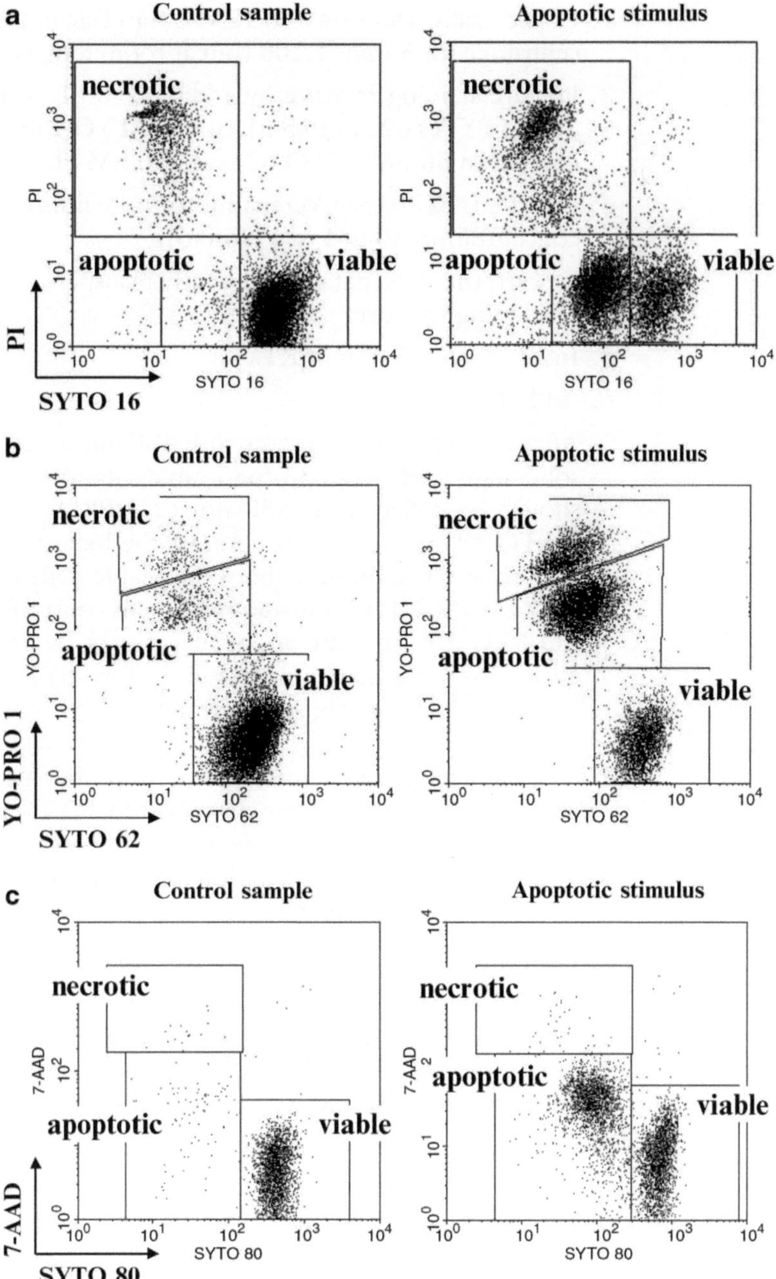

Fig. 1. Discrimination of viable, apoptotic, and late apoptotic/necrotic cells using selected SYTO probes and plasma membrane permeability markers: (**a**) SYTO 16 probe in conjunction with plasma membrane permeability marker propidium iodide (PI). Both probes were excited using 488 nm laser. SYTO 16 and PI fluorescence signals were logarithmically amplified using 530 and 575 nm band-pass filters, respectively. Debris were excluded electronically. Analysis based on *bivariate dot plots* SYTO 16 vs. PI is shown. (**b**) SYTO 62 probe in conjunction with plasma membrane permeability marker YO-PRO 1. Probes were excited using 633 and 488 nm lasers, respectively. Logarithmically amplified fluorescence signals were collected using 660 and 530 nm band-pass filters, respectively. Debris were excluded electronically. (**c**) SYTO 80 probe in conjunction with plasma membrane permeability marker 7-aminoactinomycin D (7-AAD). Both probes were excited using 488 nm laser. SYTO 80 and 7-AAD fluorescence signals were logarithmically amplified using 575 and 677 nm band-pass filters, respectively. Debris were excluded electronically.

1. Collect cell suspension into 12×75 mm Falcon FACS tube and centrifuge for 5 min, 1,100 rpm at room temperature (RT).
2. Prepare staining mixture by adding 970 μL of PBS, 5 μL of 10 μM SYTO 62, and 25 μL of 10 μM YO-PRO 1 (final concentration 50 nM SYTO 62 and 250 nM YO-PRO 1).
3. Add 1 μL of 30 mM Verapamil to the staining mixture (final concentration 30 μM; see Note 6).
4. Discard the supernatant and gently resuspend cells in 100 μL of staining mixture.
5. Incubate for 15 min at RT.
6. Add 500 μL of PBS.
7. Analyze on a flow cytometer with 488 nm (Argon-ion laser or solid-state laser) and 633/635 nm excitation lines. Emissions should be collected at 530 nm (YO-PRO 1) and 660 nm (SYTO 62) (see Note 10). Adjust the logarithmic amplification scale to distinguish between viable cells (bright SYTO 62^{+}/YO-PRO 1^{-}), early apoptotic cells (dim SYTO 62^{+}/dim YO-PRO 1^{+}), and late apoptotic and/or necrotic cells with compromised plasma membranes (low SYTO 62^{+}/bright YO-PRO 1^{+}) (Fig. 1b; see Note 10).

3.3. Detection of Apoptosis Using SYTO 80 and 7-AAD

This assay utilizes a new orange fluorescent probe SYTO 80, and its principle is similar to the above described SYTO 16 and SYTO 62 protocols (see Notes 1–3) (4). The main advantage of this protocol is that channels from other excitation lines, e.g., 633 or 405 nm are free to combine this protocol with, e.g., immunophenotyping.

1. Collect cell suspension into 12×75-mm Falcon FACS tube and centrifuge for 5 min, 1,100 rpm at room temperature (RT).
2. Prepare staining mixture by adding 875 μL of PBS, 25 μL of 10 μM SYTO 80, and 100 μL of 50 μg/mL 7-AAD (final concentration 250 nM SYTO 80 and 5 μg/mL 7-AAD).
3. Add 1 μL of 30 mM Verapamil to the staining mixture (final concentration 30 μM; see Note 6).
4. Discard the supernatant and gently resuspend cells in 100 μL of staining mixture.
5. Incubate for 15 min at RT.
6. Add 500 μL of PBS.
7. Analyze on flow cytometer with 488 nm excitation line (Argon-ion laser or solid-state laser) with emissions collected at 575–610 nm (SYTO 80) and ≥670 nm (7-AAD). Adjust the logarithmic amplification scale to distinguish between viable cells (bright SYTO 80^{+}/7-AAD^{-}), early apoptotic cells

(dim SYTO 80^{+}/dim 7-AAD^{+}), and late apoptotic and/or necrotic cells with compromised plasma membranes (low SYTO 80^{+}/bright 7-AAD^{+}) (Fig. 1c).

4. Notes

1. The universal term "apoptosis," has a propensity to misinterpret the actual phenotype of cell suicide program (1, 11–13). Thus, the use of the generic term apoptosis should be always accompanied by listing the particular morphological and/or biochemical apoptosis-associated feature(s) that was (were) detected (1–3).
2. Morphological criteria (examined by the light, fluorescent, and electron microscopy) are still the "gold standard" to define the mode of cell death and confirm results obtained by flow cytometry (1–4). Lack of microscopic examination may potentially lead to the misclassification and false positive or negative artifacts, and is a common drawback of the experimental design (1–4). The best example of such misclassification is identification of phagocytes that engulfed apoptotic bodies as individual apoptotic cells (3).
3. Following initiation of caspase-dependent apoptosis, cells loaded with selected SYTO stains exhibit gradual reduction in fluorescence signal intensity to dim values. This phenomenon substantially precedes plasma membrane permeability changes (4, 6, 7). Evidence from recently published data indicates an overall higher sensitivity of SYTO probes in the detection of early apoptotic events as compared to Annexin V-based assays (4, 7, 8). When progression toward the terminal stages of cellular demise advances, loss of SYTO fluorescence intensifies, and this usually coincides with the increased plasma membrane permeability to PI and 7-AAD (4, 7, 8).
4. We have recently shown that SYTO 16 allows discrimination between primary and secondary necrotic cells (7). Therefore, SYTO 16 provides substantial enhancement over the standard PI exclusion assay in discerning cell demise mode by flow cytometry (7).
5. Importantly, SYTO probes prove in many instances inert and safe for tracking cells over extended periods of time which may open up new opportunities for single-cell analysis protocols by both fluorescent activated cell sorting (FACS) and Lab-on-a-Chip platforms (9).
6. Recent noteworthy reports provided strong evidence that at least some SYTO probes can be substrates for MDR efflux

pumps (e.g., P-glycoprotein; P-gp) (14, 15). Caution should be, thus, exercised when using SYTO probes in cells with active ABC-class transporters. It is always advisable to confirm MDR status of studied cell population. In cells with active P-gp, its inhibition (e.g., by Verapamil hydrochloride, PSC833, Cyclosporin A) is required to avoid masking of apoptotic SYTO^{dim} subpopulation by SYTO^{dim} subpopulation engendered by an active dye efflux (14, 15). Truly apoptotic reduction of SYTO fluorescence to dim values is not affected by the presence of P-gp inhibitors (14).

7. Both fluorophores are easily excited by 488 nm lasers. While propidium iodide can be detected in both 575–610 and >650 nm channels, SYTO 16 is detected using standard fluoroscein (FITC) band-pass filter at 530 nm. Due to the significant spectral overlap of SYTO16 fluorescence in 575–610 nm channels, special attention is required to apply proper compensation between both channels. Adjust the logarithmic amplification scale to distinguish between viable cells (bright SYTO 16^+/PI^- events), apoptotic cells (dim SYTO 16^+/PI^- events), and late apoptotic/necrotic cells with compromised plasma membranes (SYTO 16^-/PI^+ events) as presented in Fig. 1a. In every cell system, there is a need to optimize concentration/incubation time of SYTO 16 probe as well as PMT voltage to achieve maximal resolution between bright, dim, and low/negative SYTO 16 events (7–9).

8. One should always bear in mind that results obtained using SYTO-based assays may vary when compared to assays detecting different cellular processes. Results acquired with SYTO probes should, therefore, never be considered conclusive without verification by independent methods (4).

9. Some apoptotic markers (e.g., loss of SYTO fluorescence to dim values; Annexin V immunoreactivity or oligonucleosomal DNA fragmentation) may not be detected in specimens challenged with divergent stimuli or microenvironmental conditions (e.g., cytokines, growth factor deprivation, heterotypic cell culture, etc.). It is always advisable to simultaneously study several markers to provide a multidimensional view of advancing apoptotic cascade (1–4). Multiparameter assays detecting several cell attributes are the most desirable solution for flow cytometric quantification of apoptosis (1–4).

10. SYTO 62 fluorescence can be conveniently detected using a standard allophycocyanin (APC) 660 nm band-pass filter. Use of SYTO 62 and YO-PRO 1 requires minimal interbeam compensation. Similar to SYTO 16, SYTO 62 dim (deemed apoptotic) events will dimly stain with YO-PRO 1 as a result of early loss in the plasma membrane function. Adjust the logarithmic amplification scale to distinguish between viable

cells (bright SYTO 62^{+}/YO-PRO 1^{-}), early apoptotic cells (dim SYTO 62^{+}/dim YO-PRO 1^{+}), and late apoptotic and/ or necrotic cells with compromised plasma membranes (low SYTO 62^{+}/bright YO-PRO 1^{+}) (8).

Acknowledgments

Supported by BBSRC, EPSRC, and Scottish Funding Council, funded under RASOR (DW); NCI CA RO1 28 704 (ZD). JS received the L'Oreal Poland-UNESCO "For Women In Science" 2007 Award. Views and opinions described in this chapter were not influenced by any conflicting commercial interests.

References

1. Darzynkiewicz, Z., Juan, G., Li, X., Gorczyca, W., Murakami, T. and Traganos, F. (1997) Cytometry in cell necrobiology: analysis of apoptosis and accidental cell death (necrosis). *Cytometry.* **27**, 1–20.
2. Darzynkiewicz, Z., Li, X. and Bedner, E. (2001) Use of flow and laser-scanning cytometry in analysis of cell death. *Methods Cell Biol.* **66**, 69–109.
3. Darzynkiewicz, Z., Huang, X., Okafuji, M. and King, M.A. (2004) Cytometric methods to detect apoptosis. *Methods Cell Biol.* **75**, 307–41.
4. Wlodkowic, D., Skommer, J. and Darzynkiewicz, Z. (2008) SYTO probes in the cytometry of tumor cell death. *Cytometry A.* **73**, 496–507.
5. Wlodkowic, D. and Darzynkiewicz, Z. (2008) Please do not disturb: destruction of chromatin structure by supravital nucleic acid probes revealed by a novel assay of DNA-histone interaction. *Cytometry A.* **73**, 877–879.
6. Frey, T. (1995) Nucleic acid dyes for detection of apoptosis in live cells. *Cytometry.* **21**, 265–74.
7. Wlodkowic, D., Skommer, J., Pelkonen, J. (2007) Towards an understanding of apoptosis detection by SYTO dyes. *Cytometry A.* **71**, 61–72.
8. Wlodkowic, D., Skommer, J., Hillier, C., Darzynkiewicz, Z. (2008) Multiparameter detection of apoptosis using red-excitable SYTO probes. *Cytometry A.* **73**, 563–569.
9. Wlodkowic, D., Skommer, J., Faley, S., Darzynkiewicz, Z., Cooper, J.M. (2009) Dynamic analysis of apoptosis using cyanine SYTO probes: from classical to microfluidic cytometry. *Exp Cell Res.* **315**, 1706–14.
10. Idziorek, T., Estaquier, J., De Bels, F., Ameisen, J.C. (1995) YOPRO-1 permits cytofluorometric analysis of programmed cell death (apoptosis) without interfering with cell viability. *J Immunol Methods.* **185**, 249–58.
11. Leist, M. and Jaattela, M. (2001) Four deaths and a funeral: from caspases to alternative mechanisms. *Nat Rev Mol Cell Biol.* **2**, 589–98.
12. Kroemer, G. and Martin, S.J. (2005) Caspase-independent cell death. *Nat Med.* **11**, 725–30.
13. Blagosklonny, M.V. (2000) Cell death beyond apoptosis. *Leukemia.* **14**, 1502–1508.
14. Schuurhuis, G.J., Muijen, M.M., Oberink, J.W., de Boer, F., Ossenkoppele, G.J., Broxterman, H.J. (2001) Large populations of non-clonogenic early apoptotic CD34-positive cells are present in frozen-thawed peripheral blood stem cell transplants. *Bone Marrow Transplant.* **27**, 487–98.
15. van der Pol, M.A., Broxterman, H.J., Westra, G., Ossenkoppele, G.J., Schuurhuis, G.J. (2003) Novel multiparameter flow cytometry assay using Syto16 for the simultaneous detection of early apoptosis and apoptosis-corrected P-glycoprotein function in clinical samples. *Cytometry B.* **55**, 14–21.

Chapter 11

Multiplexing Cell Viability Assays

Helga H.J. Gerets, Stéphane Dhalluin, and Franck A. Atienzar

Abstract

Today, obtaining mechanistic insights into biological, toxicological, and pathological processes is of upmost importance. Researchers aim to obtain as many as possible data from one cell sample to understand the biological processes under study. Multiplexing, which is the ability to gather more than one set of data from the same sample, fulfills completely this objective. Obviously, multiplexing has several advantages compared to single plex experiments and probably the most important one is that data on various parameters at exactly the same time point on the same cells or group of cells can be obtained and consequently this may contribute to saving time and effort and a reduction of the costs.

In this chapter, different endpoints were measured starting from two-seeded multiwell plates, namely, cell viability, caspase-3/7 activity, lactate dehydrogenase (LDH), adenosine triphosphate (ATP), aspartate aminotransferase (AST), and glutamate dehydrogenase (GLDH) measurements. These different endpoints were analyzed together to determine the cytotoxic properties of pharmaceutical compounds and/or reference compounds. A 96-well plate was designed to allow appropriate measurement of five doses of a compound in triplicate to determine the effect of the compound on the six different endpoints. The first four endpoints (cell viability, caspase-3/7 activity, LDH, and ATP) are discussed in detail in this chapter. AST and GLDH measurements are not discussed in detail as these are fully automatic measurements and thus behind the scope of this chapter.

As an illustrating example, the reference compound tamoxifen was used to evaluate its cytotoxic properties using the hepatocellular carcinoma cell line HepG2 cells.

Key words: Cytotoxicity, Multiplexing, Screening, Cell based

1. Introduction

Multiplex recording implicates that different endpoints are measured within the same sample. A multiparametric approach has several advantages compared to single plex experiments. First of all, since cells originate from the same subculture, differences between cultures within one experiment can be excluded, reducing test-to-test variability. Secondly, several endpoints are measured

Martin J. Stoddart (ed.), *Mammalian Cell Viability: Methods and Protocols*, Methods in Molecular Biology, vol. 740,
DOI 10.1007/978-1-61779-108-6_11,

on the same cells or group of cells which lead to lower consumption of test compound, reduced handling time, and consequently lead to a lower cost. Lowering compound consumption is very important for the pharmaceutical industry because during early stages of drug development (lead generation and lead optimization) scarce amount of compound is available. Thirdly, the predictivity of combined tests is better than in individual assays (1), and fourthly, measuring only one parameter can be misleading (2). In other words, the chance of detecting a toxic compound is far greater if different parameters are measured at the same time compared to measuring only one parameter per assay.

In this chapter, cell viability assays were multiplexed with different endpoints, such as caspase-3/7 activity, lactate dehydrogenase (LDH), adenosine triphosphate (ATP), aspartate aminotransferase (AST), and glutamate dehydrogenase (GLDH) measurements. To measure all these endpoints, several kits of Promega (Promega Corporation, Madison, WI) were used, in detail; the CellTiter-Blue® Cell Viability Assay for cell viability testing, the CellTiter-Glo® Luminescent Cell Viability Assay for ATP measurement, the CytoTox-One™ Homogeneous Membrane Integrity Assay for LDH measurement, and the Apo-One® Homogeneous Caspase-3/7 Assay for caspase activity. To measure AST and GLDH contents, kits of Roche (Roche Diagnostics GmbH, Mannheim, Germany) were used. All these measurements and activities were measured in only two identical seeded 96-well plates. To determine cell mortality, the LC50 value (i.e., the concentration killing 50% of the cells) was calculated from the cell viability assay, which is based on the reduction of resazurin into a fluorescent end product, resorufin. To improve the reliability of this LC50 value, the release of ATP, as an estimation of the number of viable cells, is also taken into account to measure the LC50 value. The release of LDH and AST was used as a marker for membrane integrity (3). To investigate the mechanism of cell death (apoptosis versus necrosis), caspase-3/7 activity was measured as a marker for apoptosis, whereas necrosis was more associated with the release of LDH and AST. GLDH, a liver-specific enzyme found exclusively in the mitochondrial matrix (4), was also measured to determine the effect of drugs on mitochondria.

Well-designed cell-based in vitro screening systems that have in vivo relevance can be used to significantly improve the compound selection process to decrease the attrition rate during drug development (2). Cytotoxicity tests, like the ones described in this chapter, can be used to rank compounds and to select the most promising candidates for further development. Nevertheless, the main limitation is that these assays are low throughput.

2. Materials

2.1. Cell Culture and Compounds

1. Human hepatocellular carcinoma (HepG2) cell line (European Collection of Cell Cultures (ECCAC), Salisbury, UK).
2. Dulbecco's modified eagle's medium (DMEM) supplemented with 10% fetal bovine serum (FBS), 50 U/ml penicillin, 50 μg/ml streptomycin, 2 mmol/l L-glutamine, and 1× non-essential amino acids (NEAA). All purchased from Bio-Whittaker Inc. (Walkersville, MD).
3. Solution of trypsin (0.25%) and 1 mM EDTA.
4. Dulbecco's phosphate-buffered saline (DPBS) (Bio-Whittaker Inc.).
5. Tamoxifen (Sigma–Aldrich, St. Louis, MO).

   The dilution series of this compound will be explained in detail.

   Six microliters of stock solution (in 25% dimethylsulfoxide (DMSO)) are added to 300 μl cells. The dilution factor is 6/306×25/100 (25% DMSO final) or 1/204. Final concentration of solvent (DMSO) is 0.49%.

   Final test concentrations in plate: 5, 10, 20, 40, and 80 μM.

   Stock solution is 80 μM×204=16,320 μM.

   MW=371.53.

   Successive dilutions in 100% DMSO: Factor 2 in 100% DMSO

   16.32 mM: 0.0606 g tamoxifen in 10 ml DMSO

   8.16 mM: 1 ml 16.32 mM stock+1 ml DMSO

   4.08 mM: 1 ml 8.16 mM stock+1 ml DMSO

   2.04 mM: 1 ml 4.08 mM stock+1 ml DMSO

   1.02 mM: 1 ml 2.04 mM stock+1 ml DMSO

   Dilute stocks in 25% DMSO

   4.08 mM: 500 μl 16.32 mM stock+1.5 ml DMEM

   2.04 mM: 500 μl 8.16 mM stock+1.5 ml DMEM

   1.02 mM: 500 μl 4.08 mM stock+1.5 ml DMEM

   510 μM: 500 μl 2.04 mM stock+1.5 ml DMEM

   260 μM: 500 μl 1.02 mM stock+1.5 ml DMEM

   Add 6 μl of each dilution in 25% DMSO to 300 μl medium to obtain the final concentrations of respectively 80, 40, 20, 10, and 5 μM in the plate.
6. DMSO (Sigma–Aldrich).
7. Triton X-100 (Sigma–Aldrich):

   Add 6 μl in 300 μl=dil. 1/51. Since solvent is water there is no need for an intermediate solution.

Successive dilutions: Factor 2 in water

0.51%: 255 μl Triton X-100 in 50 ml water

0.26%: 1 ml 0.51% stock + 1 ml water

0.13%: 1 ml 0.26% stock + 1 ml water

0.064%: 1 ml 0.13% stock + 1 ml water.

Add 6 μl of each dilution to 300 μl medium to obtain the final concentrations of respectively 0.01, 0.005, 0.0025, and 0.00125% in the plate. These percentages correspond to respectively 171.4, 85.7, 42.9, and 21.4 μM.

Final test concentrations in plate: 0.01, 0.005, 0.0025, and 0.00125%.

Stock solution is 0.01% × 51 = 0.51%.

2.2. Cytotoxicity Parameters

1. CellTiter-Blue® Cell Viability Assay (Promega):
   (a) Kit content: CellTiter-Blue® reagent.
   (b) Reagent preparation: The CellTiter-Blue® is ready for use.
2. CellTiter-Glo® Luminescent Cell Viability Assay (Promega):
   (a) Kit content: CellTiter-Glo® Buffer and CellTiter-Glo® Substrate (lyophilized).
   (b) Reagent preparation:
      - Thaw and equilibrate the CellTiter-Glo® Buffer and Substrate to room temperature prior to use.
      - Prepare the CellTiter-Glo® reagent by transferring the entire volume of CellTiter-Glo® Buffer into CellTiter-Glo® Substrate vial.
      - Mix by swirling.
3. CytoTox-One™ Homogeneous Membrane Integrity Assay (Promega):
   (a) Kit content: Substrate mix (diaphorase, lactate, NAD+), assay buffer, lysis solution and stop solution.
   (b) Reagent preparation:
      - Thaw and equilibrate the substrate mix and the assay buffer to room temperature prior to use.
      - Prepare the CytoTox-One™ reagent by adding 11 ml of assay buffer to a vial of substrate mix.
      - Mix by swirling.
4. Apo-One® Homogeneous Caspase-3/7 Assay (Promega):
   (a) Kit content: 100× Caspase Substrate Z-DEVD-R110 and Apo-One® Homogeneous Caspase-3/7 Buffer.
   (b) Reagent preparation:

- Thaw and equilibrate the Caspase Substrate to room temperature prior to use.
- Prepare the Apo-One® Caspase-3/7 reagent by adding the substrate to the buffer (ratio of 1:100).
- Mix.

5. AST (Roche Diagnostics):
   (a) Kit reference number: 11876848
6. GLDH (Roche Diagnostics):
   (a) Kit reference number: 1929992

3. Methods

A schematic overview of the plate design and the multiplexing strategy is presented in Fig. 1.

3.1. Cell Culture

Any kind of cell line can be used to seed the plates, however, make sure that the cells have the potential to stick to the bottom of the wells or that a coating is used to stick the cells to the bottom of the plate. In the example shown in this chapter, the human carcinoma cell line (HepG2) was used, therefore the handling of this cell line is explained in detail.

1. Prepare the DMEM culture medium.
2. Prewarm the medium and the trypsin–EDTA solution to 37°C.
3. Aspirate the medium from the incubation flask and rinse with 10 ml DPBS.
4. Aspirate the DPBS, make sure that no DPBS is left. If any DPBS remains in the incubation flask this can interfere with the trypsin–EDTA solution treatment, which leads to an insufficient detaching of the cells.
5. Add 5 ml trypsin–EDTA to the cells and incubate for 2–5 min at 37°C (longer incubations could be detrimental for the cells).
6. Shake vigorously the plates so that cells can detach.
7. Add 5 ml medium to the flasks and try to recover as many cells as possible.
8. Transfer the medium containing the cells to a 50-ml Falcon tube.
9. Rinse the flask with another 5 ml medium and add this into the same Falcon.
10. Centrifuge the cells at 1,100 × g for 10 min at RT.

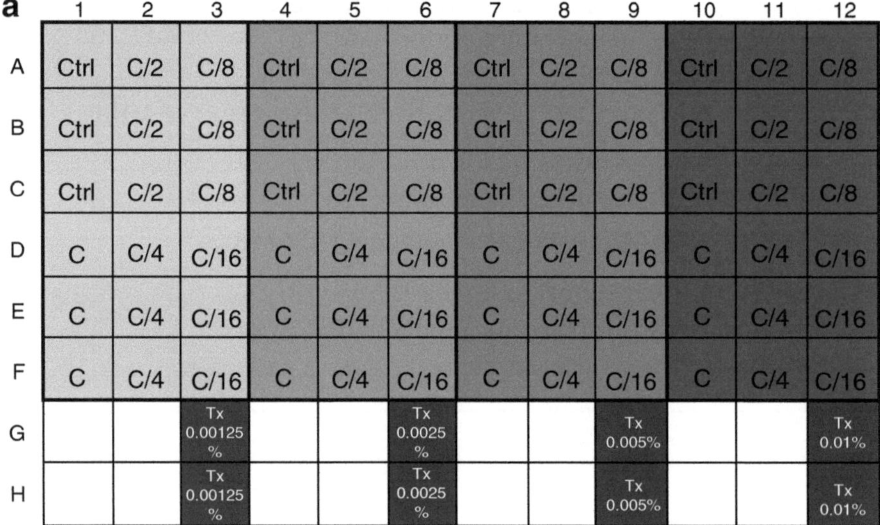

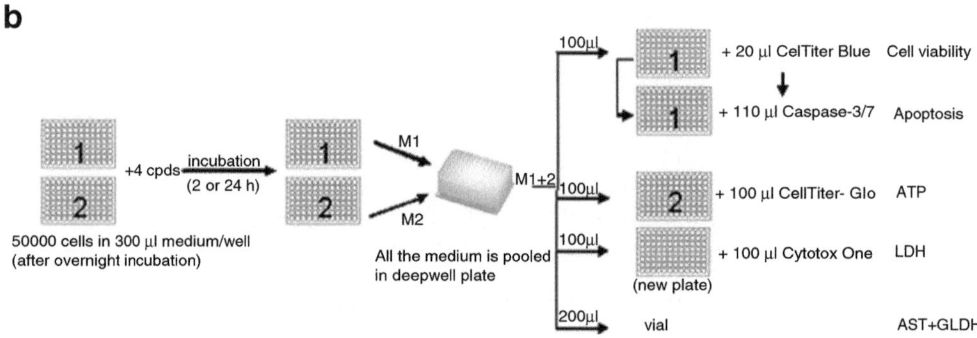

Fig. 1. Schematic overview of the plate design (**a**) and our multiplexing strategy (**b**). (**a**) Four different compounds were tested on each plate (indicated with four different *gray* tones). Each compound was tested in five different concentrations (**c**) in a 1/2-dilution series. Positive controls and wells without cells (*blank wells*) were also included in the design. Triton X-100 (Tx) was used (wells in *black*) as the positive control in a ½ dilution series. (**b**) Two 96-well plates per time point were filled with 50,000 HepG2 cells in 300 µl medium per well and allowed to attach overnight. The next day the compounds were added to the plate, each concentration was done in triplicate (technical replicates), and the cells were incubated for 2 or 24 h. Afterward, all the medium was collected in a deep well plate (in total 600 µl medium): 100 µl returned to plate 1 and was used for cell viability testing; the same plate was later used to measure caspase activity; 100 µl medium returned to plate 2 and was used to measure ATP content; 100 µl of medium was added to a new plate and was used to measure the release of LDH. Finally, 200 µl was transferred to a vial and was used to measure AST and GLDH parameters. Each experiment was repeated three different times (biological replicates). *M* medium (reprinted from (2) with permission from Elsevier).

11. Gently aspirate the supernatant and resuspend the pellet well in 25 ml medium (a brief vortex can be used but longer vortexing could damage the cells).
12. Count the cells using a cell counter and calculate the number of cells to be added to a new flask (175 cm^2) until the next passage (see Note 1).
    (a) For 2 days: take 50,000 cells/cm^2⇒8.75 million cells per flask [amount of cells needed is 175 × 50,000 cells, i.e., 8.75 million cells (in ±60 ml of medium)]

(b) For 3 days: take 30,000 cells/cm^2⇒5.25 million cells per flask
(c) For 4 days: take 20,000 cells/cm^2⇒3.50 million cells per flask
(d) For 5 days: take 15,000 cells/cm^2⇒2.63 million cells per flask

3.2. Solvent

Most compounds are not soluble in water, therefore DMSO is a good alternative to dissolve the compounds. Because cells are susceptible to DMSO the final amount of DMSO in the well should not be higher than 1%, however using only 0.5% DMSO is preferable (according to our data) and 0.1% is optimal to minimize the effect of DMSO (2). Ideally, prepare a stock solution of the compound in 100% DMSO and make intermediate dilutions in culture medium to obtain a final concentration of maximum 0.5% DMSO in the plate, or lower. This is explained in detail in Subheading 2 for tamoxifen.

3.3. Controls

As a positive control, a 1/2-dilution series of Triton X-100 was used, as explained in detail in Subheading 2. Final concentrations of Triton X-100 in the plate were 10.7, 21.4, 42.9, 85.7, and 171.4 μM. Wells without cells were also used.

3.4. CellTiter-Blue® Cell Viability Assay Multiplexed with the Apo-One® Homogeneous Caspase-3/7 Assay

1. Seed two 96-well plates with an appropriate number of cells in 300 μl medium per well and culture overnight at 37°C in a humidified 5% CO_2 and 95% air atmosphere to equilibrate and allow for attachment.
2. Treat the cells with the drug of interest for a certain time period. Because of solubility problems, most drugs should be dissolved in DMSO, make sure that the final concentration of DMSO in the well is not higher than 0.5%. Depending on the cell type used, a final DMSO concentration of 1% or more could damage the cells particularly after long-term incubation.
3. After the drug incubation period, collect the medium from the first plate to a deep 96-well plate and directly re-add 100 μl of the conditioned medium back to the cells (see Note 2). Do not throw the remaining medium as it will be used for other endpoint measurements. The medium can be stored in the deep well plate in the CO_2-incubator and should be used within 2–3 h.
4. Add 20 μl CellTiter-Blue® reagent to each well and gently mix to ensure that the reagent is equally dispersed. Incubate the plate for 1 h at 37°C in a humidified 5% CO_2 and 95% air atmosphere (see Note 3).
5. Record fluorescence ($560_{EX}/590_{EM}$) to measure cell viability.

6. Re-use the plate and add 120 μl Apo-One Caspase reagent to each well and incubate the plate for an additional hour at room temperature under constant agitation. This will lyse the cells. The volume of Apo-One Caspase reagent added to each well should give a 1:1 proportion with the CellTiter-Blue reagent volume already in the plate. The incubation period can be extended up to 2 h but still under constant agitation (see Note 4).
7. Record fluorescence ($560_{EX}/590_{EM}$) to measure the caspase activity.

For more details of these two assays consult Promega Technical Bulletin #317 and #295 that describes the CellTiter-Blue and the Caspase-3/7 assays, respectively.

3.5. CellTiter-Glo® Luminescent Cell Viability Assay Multiplexed with the CytoTox-One™ Homogeneous Membrane Integrity Assay

1. Seed a 96-well plate with an appropriate number of cells in 300 μl medium per well and culture overnight at 37°C in a humidified 5% CO_2 and 95% air atmosphere to equilibrate and allow for attachment.
2. Treat the cells with the drug of interest. The time should be the same as in step 2 of Subheading 3.4.
3. After the drug incubation period collect the medium from the second plate to the same deep 96-well plate as used under Subheading 3.4 (the medium will be mixed with the previously collected medium) and directly re-add 100 μl of the conditioned medium back to the cells, this plate with cells will be used to measure the CellTiter-Glo cell viability assay. Part of the collected medium in the deep 96-well plate will be used to measure the CytoTox-One assay, as for this assay no cells are necessary, only the medium.

3.5.1. CellTiter-Glo Assay

1. Add 100 μl CellTiter-Glo reagent to each well containing cells and gently shake the plate for 10 min at room temperature.
2. Record the luminescence.

3.5.2. CytoTox-One Assay

1. Fill a new, sterile 96-well plate with 100 μl of medium, taken from the deep well plate, to each well.
2. Add 100 μl of CytoTox-One reagent to each well and incubate the plate for 10 min at room temperature without agitation.
3. Stop the reaction by adding 50 μl of stop solution to each well.
4. Record the fluorescence ($485_{EX}/530_{EM}$).

For more details of these two assays consult Promega Technical Bulletin #288 and #306 that describes the CellTiter-Glo and the CytoTox-One assays, respectively.

3.6. Remaining Supernatant

The supernatant remaining in the deep 96-well plate, which is between 200 and 300 μl, can be used to measure other parameters, like AST and GLDH. For these particular endpoints, the remaining medium was added to a vial and put into the Hitachi analyzer Modular P of Roche-Diagnostics GmbH (Mannheim, Germany). The protocols to measure these endpoints are not discussed in detail in this chapter, because the measurements of these endpoints were performed on a robot and thus behind the scope of this chapter. Obviously, other parameters such as enzymatic activities of alanine aminotransferase (ALT) and alkaline phosphatase (ALP) can also be measured on the same supernatant using the Hitachi analyzer.

Figure 2 shows an example of the measurement of four different endpoints when increasing amounts of tamoxifen were added to the HepG2 cells.

In order to visualize the correlations between the six different parameters and to study the consistency between the replicates, a principal component analysis (PCA) was carried out as shown in Fig. 3 for tamoxifen. PCA is a statistical technique used to reduce the dimensionality of a dataset without much loss of information. The outputs of PCA are maps of two types: correlation circles and scatter plots of scores. The correlation circles are maps which show the correlations between the variables. The scatter plots of scores are the projections of the cloud of data points in the

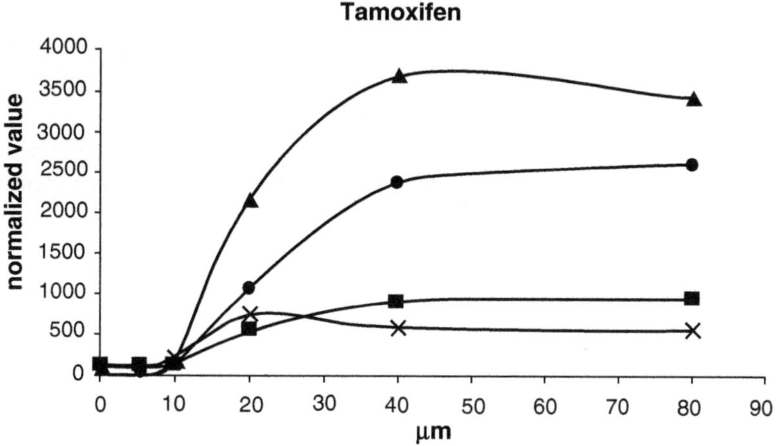

Fig. 2. Effect of tamoxifen exposure on HepG2 cells for 24 h on caspase, LDH, AST, and GLDH measurements. HepG2 cells were cultured overnight at 50,000 cells per well in a 96-well plate. The cells were exposed to tamoxifen at different concentrations as indicated in the graph. Parameters measured were LDH (*filled triangle*), caspase (*cross sign*), AST (*filled square*), and GLDH (*filled circle*). The final DMSO concentration in the wells was 0.5%. For each parameter, the mean value of the control wells was set to 100%. The mean value for the other concentrations was calculated proportionally to this control value. Each experiment has been repeated three times (biological replicates) and the mean value is represented in the graphs (reprinted from (2) with permission from Elsevier).

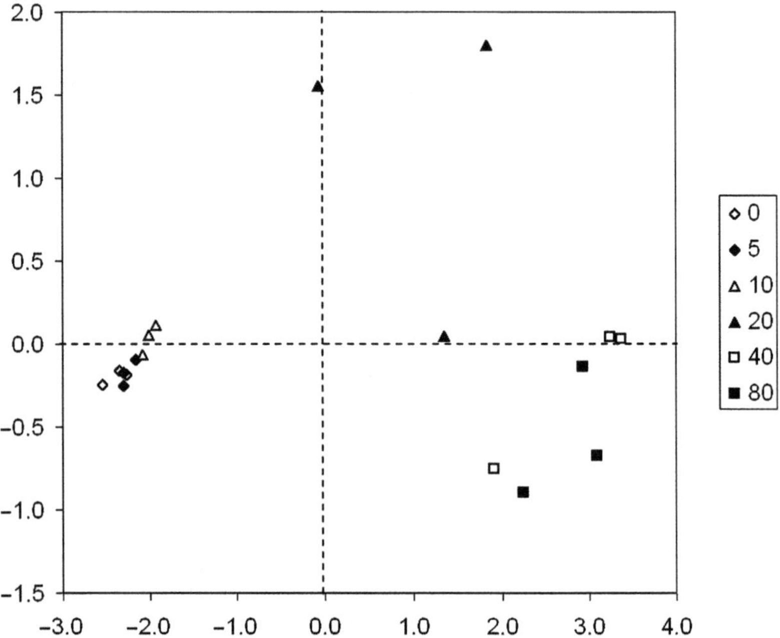

Fig. 3. Principal component analysis (PCA) of tamoxifen after 24 h of incubation. HepG2 cells were exposed to different concentrations of tamoxifen (0, 5, 10, 20, 40, and 80 μM) for 24 h. Afterward the cell viability, ATP, LDH, AST, GLDH, and caspase activity were measured. A PCA plot of each concentration of the compound was created by integrating the six different parameters. More information concerning PCA plot is explained in the text (reprinted from (2) with permission from Elsevier).

multidimensional space on the plane that show the maximum information, i.e., the maximum variance. The relationships between the parameters (*r*) were mainly studied on the circle of correlation issued from a PCA carried out on the results of the maximal concentration of each compound.

4. Notes

1. Before the start of the experiment, a dilution series of the cells should be seeded into a 96-well plate and the cell viability should be measured in order to determine the optimal number of cells per well. It is important that the cells at the density seeded can still proliferate and that they are not at confluence. For the hepatocellular carcinoma cell line (HepG2 cells), the optimal number of cells was found to be 50,000 cells per well (2).
2. If cells were directly incubated in 100 μl of medium there is a risk that some of the medium evaporates overnight and this

would lead to an incorrect proportion between the amount of medium and the amount of reagent added.

3. An incubation period of 1 h with the CellTiter-Blue® reagent gave good result using the HepG2 cell line. However, the incubation period can be extended up to 4 h, depending on the cell line and the number of cells used and therefore this should be determined empirically.
4. Always follow a sequential order of assay chemistry addition when multiplexing a viability assay with a caspase activity assay. This is required because homogeneous caspase assay formulations contain agents that lyse cells (5).

References

1. O'Brien, P.J., Irwin, W., Diaz, D., Howard-Cofield, E., Krejsa, C.M., Slaughter, M.R., Gao, B., Kaludercic, N., Angeline, A., Bernardi, P., Brain, P. and Hougham, C. (2006). High concordance of drug-induced human hepatotoxicity with in vitro cytotoxicity measured in a novel cell-based model using high content screening. Arch Toxicology, **80**, 580–604.
2. Gerets, H.J.J., Hanon, E., Cornet, M., Dhalluin, S., Depelchin, O., Canning, M., and Atienzar, F.A. (2009). Selection of cytotoxicity markers for the screening of new chemical entities in a pharmaceutical context: A preliminary study using a multiplexing approach. Toxicology In Vitro, **23**, 319–332.
3. Riss, T.L., and Moravec, R.A. (2004). Use of multiple assay endpoints to investigate the effects of incubation time, dose of toxin, and plating density in cell-based cytotoxicity assays. ASSAY and Drug Development Technologies, **2**, 51–62.
4. O'Brien, P.J., Slaughter, M.R., Polley, S.R. and Kramer, K. (2002). Advantages of glutamate dehydrogenase as a blood biomarker of acute hepatic injury in rats. Laboratory Animals, **36**, 313–321.
5. Niles, A.L., Moravec, R.A. and Riss, T.L. (2008). Multiplex caspase activity and cytotoxicity assays. Methods Molecular Biology, **414**, 151–162.

Chapter 12

Cytotoxicity Testing: Measuring Viable Cells, Dead Cells, and Detecting Mechanism of Cell Death

Terry L. Riss, Richard A. Moravec, and Andrew L. Niles

Abstract

Testing the effects of compounds on the viability of cells grown in culture is widely used as a predictor of potential toxic effects in whole animals. Among the several alternative assays available, measuring the levels of ATP is the most sensitive, reliable, and convenient method for monitoring active cell metabolism. However, recently developed combinations of methods have made it possible to collect more information from in vitro cytotoxicity assays using standard fluorescence and luminescence plate readers. This chapter describes two assay methods. The first utilizes beetle luciferase for measuring the levels of ATP as a marker of viable cells. The second more recently developed multiplex method relies on selective measurement of three different protease activities as markers for viable, necrotic, and apoptotic cells. Data analysis from the measurement of three marker protease activities from the same sample provides a useful tool to help uncover the mechanism of cell death and can serve as an internal control to help identify assay artifacts.

Key words: Cell viability, Cytotoxicity, Apoptosis, Multiplexing, Luciferase, Luminescence, Fluorescence, Membrane integrity, ATP

1. Introduction

In vitro cytotoxicity assays measure whether a test compound is toxic to cells in culture, usually by determining the number of viable cells remaining after a defined incubation period. The desired approach is to use a convenient and cost-effective method that predicts in vivo toxicity by measuring a surrogate marker to indicate the viable cell number compared to untreated controls. The methods of estimating the number of viable cells in culture are usually based on measuring an indicator of metabolic activity. Several methods have been developed to measure metabolically active viable cells based on their ability to convert certain classes of chemicals into forms that can be easily measured.

Martin J. Stoddart (ed.), *Mammalian Cell Viability: Methods and Protocols*, Methods in Molecular Biology, vol. 740,
DOI 10.1007/978-1-61779-108-6_12, © Springer Science+Business Media, LLC 2011

The MTT tetrazolium assay was the first convenient 96-well method developed for screening large numbers of samples (1). The MTT tetrazolium compound is reduced by viable cells into an intensely colored formazan precipitate that subsequently is solubilized into a uniformly colored solution with a second procedural step before 570-nm absorbance is measured using a plate reading spectrophotometer. The colored product is directly proportional to the number of viable cells.

Improved tetrazolium reagent chemistries (e.g., MTS, XTT, WST) were developed that could be converted by cells directly into aqueous soluble formazans, thus eliminating the solubilization step required for the MTT reagent (2–4). The improved tetrazolium reagents provided the first convenient "add–incubate–read" method of measuring viable cell number. Although the sensitivity of detection is limited to approximately 1,000 viable cells, these assays have been adopted by many cell biology laboratories but not by automated HTS laboratories.

Resazurin is a similar redox-sensitive dye that is reduced by viable cells into a fluorescent compound, resorufin. The homogeneous add–incubate–measure resazurin reduction assay provides greater sensitivity of detection than tetrazolium methods and has been used successfully for high-throughput screening (5, 6). However, the resazurin assay is hampered by limitations caused by the intense blue color of the resazurin solution and the occurrence of fluorescence interference of test compounds with the resorufin product.

The major disadvantages of the tetrazolium and resazurin reduction assays are that they require a 1–4-h period of incubation with viable cells for conversion into the colorimetric or fluorescent indicator; chemical interference is caused by reducing compounds; and perhaps most importantly, the tetrazolium and resazurin compounds used to indicate cell viability are themselves toxic to cells.

Detection of ATP has become the gold standard cell viability assay for high-throughput screening. The level of ATP is closely regulated by viable cells. The process of cell death results in a loss of ability to synthesize new ATP along with a rapid depletion of cytoplasmic ATP by endogenous ATPases. These properties combine to make ATP a valid marker of viable cells in culture (7, 8). Measurement of ATP is accomplished by taking advantage of the properties of the firefly luciferase enzyme to generate a luminescent signal (9).

Considerable effort has been devoted to optimize reagent chemistries to make the modern version of the ATP assay the most sensitive and the fastest method available for plate-based cell viability assays. Assay sensitivity is achieved because of the extremely low background luminescence which enables large signal to background ratios. Assay speed and convenience are achieved because a homogeneous single reagent addition protocol results in immediate cell lysis and generation of a glowing luminescent signal. Unlike the tetrazolium or resazurin reduction assays, the need to co-incubate viable cells with assay reagent for hours is eliminated, resulting in lowering the probability of artifacts.

The most significant advancement in ATP assay technology resulted from the use of directed evolution to create luciferase mutants that retain enzymatic activity in reagent formulations containing ATPase inhibitors and harsh detergents to lyse cells (10). The stable form of luciferase has been demonstrated to be far less prone to artifacts caused by interference of compounds in chemical libraries (11). The possibility of artifacts resulting from luciferase inhibitors still needs to be addressed, but can be overcome by counter screening using a different assay endpoint or by using multiplexing techniques with more than one assay chemistry to measure different markers.

One recently developed method of multiplexing involves the measurement of separate protease activities as markers of viable and dead cells (12). A protease activity found predominantly in intact viable cells is measured using a fluorogenic cell permeable peptide substrate (glycyl-phenylalanyl-aminofluorocoumarin; GF-AFC). The GF-AFC substrate permeates intact cells where it is cleaved to generate a fluorescent signal proportional to the number of living cells. Upon cell death, the protease that cleaves the GF-AFC substrate becomes inactive. A second protease activity released into the medium from dead cells that have lost membrane integrity is measured using a nonpermeable fluorogenic peptide substrate (bis-alanylalanyl-phenylalanyl-rhodamine 110; bis-AAF-R110). Because the AAF-R110 substrate is nonpermeable and cannot enter viable cells, the signal from this substrate is selective for the dead cell population. The two substrates can be combined and added as a single reagent to cells in culture and incubated for 30 min to generate fluorescent signals proportional to the number of viable and dead cells. The viable and dead cell protease markers show excellent correlation with established methods including measurement of ATP and release of lactate dehydrogenase (13). The GF-AFC and AAF-R110 substrates are compatible with other reagents to enable additional downstream multiplexing to measure other parameters, such as a luminescent reporter assay to measure gene expression or a luminescent caspase assay as a marker of apoptosis.

Cells undergoing programmed cell death (apoptosis) in vitro express various transient markers of the process and initially maintain an intact cell membrane; however, all cells undergoing apoptosis in vitro eventually undergo secondary necrosis to release cytoplasmic components into the culture medium and thus would be scored as dead/necrotic using a protease release assay. The apoptotic cells present in a population can be detected by measuring caspase-3 activity. The caspase-3 executioner protease is measured using a Z-DEVD-aminoluciferin substrate that is cleaved by active caspase to release aminoluciferin that can be detected using a beetle luciferase reaction to generate a luminescent signal. Information regarding whether or not cells die via an apoptotic mechanism is valuable for some cytotoxicity studies. A method to multiplex the measurement of the number of viable, necrotic, and apoptotic cells in the same sample in multiwall plates is described in the protocols given below. Data analysis from the measurement of three marker protease activities from the same sample provides a useful tool to help uncover the mechanism of cell death and can serve as internal control to help identify assay artifacts.

2. Materials

1. Human K562 and murine L929 cells (American Type Culture Collection, Manassas, VA) were maintained in stock culture using standard cell culture techniques.
2. RPMI 1640 medium supplemented with 10% fetal bovine serum (FBS).
3. Opaque-walled multiwell plates adequate for cell culture (see Note 1).
4. Multichannel pipette (Rainin Instrument, Oakland, CA) or similar automated pipetting station adequate for reagent delivery into multiwell plates.
5. Reagent reservoirs compatible with multichannel pipettes.
6. ATP detection reagent (CellTiter-Glo Assay, Promega Corporation, Madison, WI).
7. Reagent for multiplex detection of viable, necrotic, and apoptotic cells (ApoTox-Glo Assay, Promega). This comprises GF-AFC substrate (100 mM in DMSO), bis-AAF-R110 substrate (100 mM in DMSO), and five bottles of Caspase-Glo® 3/7 substrate (lyophilized).
8. 17-(Allylamino)-17-demethoxygeldanamycin (17-AAG, Enzo Life Sciences International, Plymouth Meeting, PA) and epoxomicin (EMD, La Jolla, CA) are dissolved in culture medium at twice the highest concentration tested and serial twofold dilutions are made in replicate wells across assay plate.
9. Plate shaker for mixing multiwell plates.
10. Circulating thermal block 2,050 W to maintain constant temperature of assay plates on the bench.

11. Microplate reader (GloMax™ Multimodal, Promega, Madison, WI or BMG LABTECH POLARstar OPTIMA, BMG LABTECH, Durham, NC) capable of measuring luminescence and fluorescence at the following sets of wavelengths: (a) excitation 400 nm and emission 505 nm and (b) excitation 485 nm and emission 520 nm.
12. GraphPad Prism™ software (GraphPad, La Jolla, CA).

3. Methods

More detailed protocols with additional background information on characterization of the assay chemistries, optional controls, and examples of applications can be found in the technical manuals describing the CellTiter-Glo® Luminescent Cell Viability Assay that measures ATP (http://www.promega.com/tbs/tb288/tb288.html) and the ApoTox-Glo™ Triplex Assay that measures viable, dead, and necrotic cells (http://www.promega.com/tbs/tm322/tm322.html).

3.1. ATP Assay for Viable Cells

3.1.1. Reagent Preparation

1. Thaw the CellTiter-Glo® Buffer, and equilibrate to room temperature prior to use. The buffer may be thawed and stored at room temperature for up to 48 h prior to use (see Note 2).
2. Equilibrate the lyophilized CellTiter-Glo® Substrate to room temperature prior to use.
3. Transfer the entire liquid volume of buffer into the amber bottle containing substrate to reconstitute the lyophilized enzyme/substrate mixture. This forms the CellTiter-Glo® Reagent containing detergent to lyse cells and luciferase and luciferin as the key ingredients for the measurement of ATP.
4. Mix by gently vortexing, swirling, or by inverting the contents to obtain a homogeneous solution. The substrate should go into solution easily in less than 1 min.

3.1.2. Protocol for the Cell Viability Assay

1. Prepare opaque-walled multiwell plates with mammalian cells in culture medium, 100 μl per well for 96-well plates or 25 μl per well for 384-well plates (see Note 3). Multiwell plates must be compatible with the luminometer used (see Note 4).
2. Prepare control wells containing culture medium without cells to obtain a value for background luminescence (see Note 5).
3. Add serial dilutions of the compound to be tested to experimental wells, and incubate at 37°C in a humidified 5% CO_2 atmosphere according to culture protocol, usually 24–48 h (see Notes 6 and 7).

4. Equilibrate the plate and its contents at room temperature for approximately 30 min (see Note 8).
5. Add a volume of CellTiter-Glo® Reagent equal to the volume of cell culture medium present in each well (e.g., add 100 μl of reagent to 100 μl of medium containing cells for a 96-well plate, or add 25 μl of reagent to 25 μl of medium containing cells for a 384-well plate) (see Note 9).
6. Mix contents for 2 min on an orbital shaker to induce cell lysis.
7. Allow the plate to incubate at room temperature for at least 10 min to stabilize luminescent signal (see Note 10).
8. Record the luminescence (see Note 11).

An example set of results from an ATP assay of L929 cells treated with various concentrations of TNFα is shown in Fig. 1. Increasing the concentrations of TNFα result in a loss of cell viability, a decrease in the amount of ATP per cell, and thus a reduction in the luminescent signal. Concentrations of TNFα above 5–10 ng/ml kill essentially all the cells using the assay conditions described.

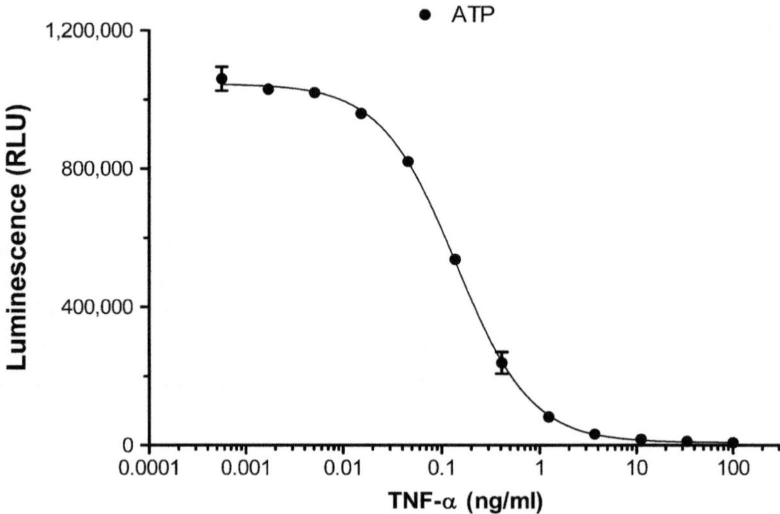

Fig. 1. Viability by ATP detection of L929 cells treated with TNFα. Murine L929 cells were plated in a 96-well plate at a density of 10,000 cells/80 μl and cultured in a humidified incubator at 37°C, 5% CO_2 for 24 h. Serial dilutions of TNFα (Promega Corporation, Madison, WI) were prepared at 5× final concentration in culture medium containing 5 μg/ml actinomycin D (Sigma, St. Louis, MO) and 20 μl added to each well. The plate was returned to 37°C, 5% CO_2 for an additional 18 h. Following a 30-min equilibration at 22°C using a circulating thermal block, an equal volume (100 μl/well) of CellTiter-Glo® Reagent was added to each well and mixed by shaking for 2 min. The plate was incubated at 22°C for 45 min prior to recording luminescence using a BMG LABTECH PolarStar plate luminometer. Four-parameter curve fit analysis was performed using GraphPad Prism software (GraphPad, San Diego, CA). The EC_{50} of TNFα for these conditions was 0.14 ng/ml. The vehicle control of zero TNF (1,082,366 ± 34,804 RLU) is not shown on the log scale.

3.2. Multiplexed Protease Assays to Measure Viable, Necrotic, and Apoptotic Cells Using the ApoTox-Glo Assay

3.2.1. Reagent Preparation

1. Thaw assay buffer, GF-AFC substrate, and bis-AAF-R110 substrate at 37°C in a water bath. Thaw Caspase-Glo® 3/7 buffer and substrate at room temperature.
2. Transfer 10 μl of each of the GF-AFC substrate and the bis-AAF-R110 substrate into 2.0 ml of assay buffer. Vortex the contents until the substrates are thoroughly dissolved. This mixture will be referred to as the viability/cytotoxicity reagent (see Note 12 for storage of reagents).
3. Transfer the contents of the Caspase-Glo® 3/7 buffer bottle into the amber bottle containing the lyophilized Caspase-Glo® 3/7 substrate (Z-DEVD-aminoluciferin), luciferase, and ATP. Mix by swirling or inverting the contents until the substrate is thoroughly dissolved to form the Caspase-Glo® 3/7 reagent (see Note 13).

3.2.2. Example Multiplex Assay Protocol for 96-Well Plate Format

1. Set up 96-well assay plates containing cells in culture medium at an appropriate density (see Note 14).
2. Add various dilutions of test compounds and vehicle controls to appropriate wells for a final volume of 100 μl per well for 96-well format.
3. Culture cells at 37°C in a humidified 5% CO_2 atmosphere for the desired test exposure period (see Note 15).
4. Add 20 μl of viability/cytotoxicity reagent containing both GF-AFC and bis-AAF-R110 substrates to all wells and mix by orbital shaking (300–500 rpm for 30 s).
5. Incubate for 30 min at 37°C (see Note 16).
6. Measure fluorescence at 400 nm excitation/505 nm emission wavelengths to determine free AFC as a marker of viable cell number and 485 nm excitation/520 nm emission to determine free R110 as the marker of the number of necrotic cells with a damaged membrane.
7. Add 100 μl of Caspase-Glo® 3/7 Reagent to all wells and mix by orbital shaking (300–500 rpm for ~30 s).
8. Incubate for 30 min at room temperature (see Note 17).
9. Measure luminescence using an integration time of 0.5–1 s to determine the release of aminoluciferin indicative of caspase-3/7 activity as a marker of apoptosis.

The correlation between ATP detection and the activity of proteases, which are able to cleave GF-AFC, as markers of cell viability is shown in Fig. 2. The overlapping curves suggest a similar responsiveness of the cells using both assays.

An example showing multiplex measurement of the three protease activity as markers of viable cells, necrotic cells, and apoptosis is shown in Fig. 3.

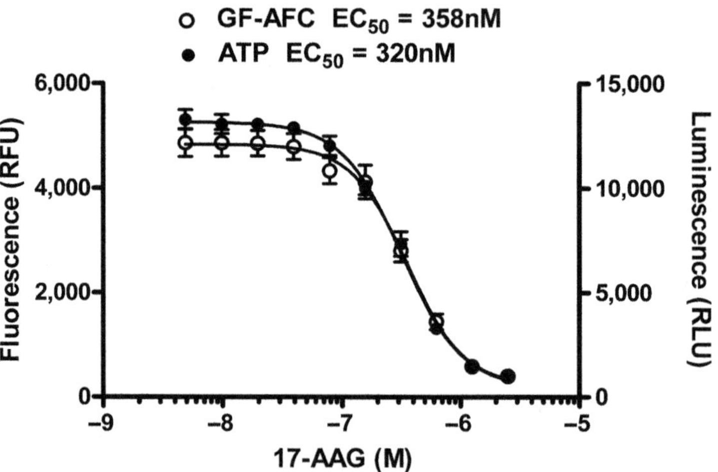

Fig. 2. 17-AAG was twofold serially diluted in RPMI 1640 + 10% FBS in 50 μl in a 96-well plate. K562 cells were added at a density of 5,000 cells/well in 50 μl. The plate was placed at 37°C in 5% CO_2 for 72 h. CellTiter-Fluor substrate (GF-AFC) and CellTiter-Glo reagents to detect ATP were prepared according to manufacturer and added in parallel wells. Fluorescence (excitation 400 nm and emission 505 nm) and luminescence were measured using a GloMax™ Multimodal instrument. Data were fitted using Prism™ software.

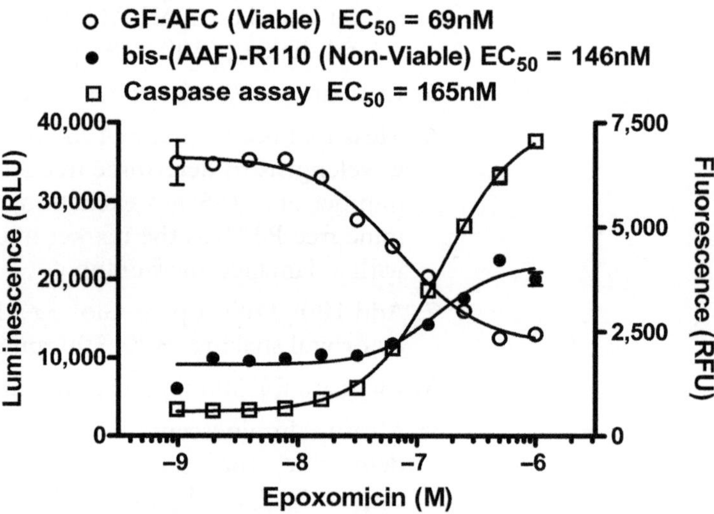

Fig. 3. Epoxomicin was twofold serially diluted in RPMI 1640 + 10% FBS in 50 μl in a 96-well plate. K562 were added at a density of 5,000 cells/well in 50 μl. The plate was placed at 37°C in 5% CO_2 for 24 h. ApoTox-Glo Reagents were prepared and added to the plate wells sequentially according to manufacturer's protocol. Fluorescence and luminescence were measured using a BMG PolarStar™ instrument. Dose–response data were fitted using Prism™ software.

4. Notes

1. Store opaque-walled multiwell plates in dark and out of direct exposure to room lighting which may contribute to background signal.
2. Temperature fluctuations affect the rate of the luciferase reaction and thus the intensity of the luminescent signal. If the room temperature fluctuates, use a constant-temperature incubator.
3. It is beyond the scope of this chapter to describe the details of cell culture; however, it is worth noting briefly that for consistent results, it is essential to follow a standard operating procedure for culturing and harvesting cells used to set up cytotoxicity assay protocols. The physiological condition of cells used to set up assays is known to affect responsiveness to toxins. Key parameters such as the density of the stock culture of cells used to set up an assay or the density of cells in the assay plates both can affect the measured IC_{50} concentration of a toxin (14). Among the most critical assay parameters to characterize during assay development is to determine the length of exposure to compounds being tested before making measurements of viability, cytotoxicity, or apoptosis. The toxic effect of compounds is a combination of concentration and duration of exposure. It is critical to understand that the dynamic changes occur in cell populations that are treated with a toxin. Those changes are dependent on the cell type and density (14). Cell line authentication also should be part of the standard operating procedures for all cell culture laboratories. The misidentification and cross-contamination of cell lines have been recognized as a significant problem (15). The NIH has posted a notice regarding the authentication of cultured cell lines (http://grants.nih.gov/grants/guide/notice-files/NOT-OD-08-017.html).
4. The number of cells to use per well will depend on several factors including the cell type. The seeding density should be empirically determined for each cell type for optimum performance. A good starting point is 10,000 cells per well in 100 µl for the 96-well format.
5. Appropriate controls are required for proper interpretation of data. It is recommended to set up a series of replicate wells on each assay plate for each of the following controls. *No-cell control* containing culture medium without cells to serve as the negative control for determining the background fluorescence and luminescence. *Untreated cells (vehicle) control* containing culture medium and untreated cells to serve as a

vehicle control. Add the same concentration and volume of solvent (e.g., DMSO or culture medium) used to deliver the test compounds. *Toxin-treated positive control* containing cells in culture medium treated with a compound known to induce toxicity or apoptosis in the cells used in your model system. *Optional test compound control* containing wells without cells containing the vehicle and test compound to test for possible interference with the assay chemistries. This test is optional and may be done in a follow-up assay only on compounds found to be active in primary screening.

6. For screening large numbers of small molecule compounds, replicates of a single concentration (usually 10 μM) instead of serial dilutions can be used. However, measuring multiple concentrations of unknown toxins using the quantitative HTS (qHTS) approach provides several advantages including the ability to identify toxins that may be missed using a single concentration (16).
7. Some cell types benefit from equilibration in the assay plates to allow attachment before the addition of test compounds. However, for some assays it may be more convenient to prepare assay plates to contain serial dilutions of test compounds prior to addition of cells. The order of those two steps should be empirically determined depending on the model system.
8. Equilibration to a constant temperature avoids artifacts caused by temperature gradients between the edge and center of the plate. This is important because the luminescent signal is dependent on the rate of the luciferase enzymatic reaction which is affected by temperature.
9. Maintaining the 1:1 ratio of culture volume to CellTiter-Glo® Reagent is important to achieve complete cell lysis and ensure a stable glow luminescence signal.
10. Uneven luminescent signal within plates can be caused by temperature gradients, uneven seeding of cells, or edge effects in multiwell plates.
11. Instrument settings depend on the manufacturer. An integration time of 0.25–1 s per well should serve as a guideline.
12. Once prepared, the viability/cytotoxicity reagent containing both substrates should be used within 24 h if stored at room temperature. Unused viability/cytotoxicity reagent can be stored at 4°C for up to 7 days with no appreciable loss of activity.
13. Reconstituted Caspase-Glo® 3/7 reagent can be stored at 4°C for up to 7 days with minimal loss of activity.
14. It is recommended to empirically determine the optimal cell number for each model system; however, a good starting point is fewer than 20,000 cells per well in a 96-well plate and fewer than 5,000 cells per well in a 384-well plate.

15. In vitro cytotoxicity is dependent upon the concentration of the toxic compound and the length of exposure to cells (14). The kinetics of cell death can vary widely between individual compounds and treatments. For example, if cells are treated with a slow-acting apoptosis-inducing compound for only 2 h, it is possible that changes in viability, cytotoxicity, or caspase activation will not be measurable, whereas longer exposure times with the same compound will likely reveal cytotoxicity. Alternatively, if cells are treated with a fast-acting toxin and exposed for a long incubation period, viability values will be low, but cytotoxicity values may be underestimated if the dead cell marker enzymatic activity diminishes after several hours of incubation in culture medium. Testing multiple concentrations will help reveal toxic effects if only a single incubation period is used for testing. It is important to characterize new compounds in multiple exposure periods (16).
16. Incubations of the GF-AFC and bis-AAF-R110 substrates longer than 30 min may improve assay sensitivity and dynamic range if low cell numbers are used. However, do not incubate more than 3 h.
17. Incubation of the Caspase-Glo® 3/7 reagent longer than 30 min may improve assay sensitivity and dynamic range. Total incubation time for the luminescent caspase assay depends upon the culture system, but typically peak luminescent signal will be reached in 1 h. For optimal results, the maximum recommended incubation time is 3 h. In general, the luminescent signal remaining at 3 h is greater than 70% of peak luminescence.

References

1. Mosmann, T. (1983) Rapid colorimetric assay for cellular growth and survival: Application to proliferation and cytotoxicity assays. *J. Immunol. Meth.* **65**, 55–63.
2. Barltrop, J. and Owen, T. (1991) 5-(3-carboxymethoxyphenyl)-2-(4,5-dimethylthiazoly)-3-(4-sulfophenyl) tetrazolium, inner salt (MTS) and related analogs of 3-(4,5-dimethylthiazolyl)-2,5-diphenyltetrazolium bromide (MTT) reducing to purple water-soluble formazans as cell-viability indicators. *Bioorg. Med. Chem. Lett.* **1**, 611–614.
3. Paull, K. D., Shoemaker, R. H., Boyd, M. R., et al., (1988) The synthesis of XTT: A new tetrazolium reagent that is bioreducible to a water-soluble formazan. *J. Heterocyclic Chem.* **25**, 911–914.
4. Ishiyama, M., Shiga, M., Sasamoto, K., Mizoguchi, M. and He, P. (1993) A new sulfonated tetrazolium salt that produces a highly water-soluble formazan dye. *Chem. Pharm. Bull. (Tokyo)* **41**, 1118–1122.
5. Ahmed, S. A., Gogal, R. M. and Walsh, J. E. (1994) A new rapid and simple nonradioactive assay to monitor and determine the proliferation of lymphocytes: An alternative to [3H]thymidine incorporation assays. *J. Immunol. Meth.* **170**, 211–224.
6. Shum, D., Radu, C., Kim, E., Cajuste, M., Shao, Y., Seshan, V. E. and Djaballah, H. (2008) A high density assay format for the detection of novel cytotoxic agents in large chemical libraries. *J. Enz. Inhib. Med. Chem.* **23(6)**, 931–945.
7. Riss T. L, Moravec R. A., O'Brien, M. A., Hawkins, E. M. and Niles, A. Homogeneous Multiwell Assays for Measuring Cell Viability, Cytotoxicity and Apoptosis. In: Minor, L. ed. Handbook of Assay Development in Drug Discovery. Taylor & Francis Group, 2006: 385–406.

8. Severson, W. E., Shindo, N., Sosa, M., Fletcher, T., White, L., Ananthan, S. and Jonsson, C.B. (2007) Development and validation of a high-throughput screen for inhibitors of SARS CoV and its application in screening of a 100,000-compound library. *J. Biomol. Screen.* **12(1)**, 33–40.
9. Lundin, A., Hasenson, M., Persson, J., and Pousette A. (1986) Estimation of biomass in growing cell lines by ATP assay. *Meth. Enzymol.* 133:27–42.
10. Hall, M. P., Gruber, M. G., Hannah, R. R., Jennens-Clough M. L., Wood K. V. (1998) Stabilization of firefly luciferase using directed evolution. In: *Bioluminescence and Chemiluminescence—Perspectives for the 21st Century*. Roda A, Pazzagli M., Kricka L.J., Stanley P.E. (eds.), pp. 392–395. John Wiley & Sons, Chichester, UK.
11. Auld, D. S., Zhang, Y-Q., Southall, N. T., Rai, G., Landsman, M., MacLure, J., Langevin, D., Thomas, C. J., Austin, C. P., and Inglese, J. (2009) A Basis for Reduced Chemical Library Inhibition of Firefly Luciferase Obtained from Directed Evolution. *J. Med. Chem.* 52(**5**), 1450–1458.
12. Niles, A. L., Moravec, R. A., Hesselberth, P. E., Scurria, M. A., Daily, W. J., and Riss, T. L. (2007) A homogeneous assay to measure live and dead cells in the same sample by detecting different protease markers. *Anal. Biochem.* **366**, 197–206.
13. Niles, A. L., Moravec, R. A., and Riss, T. L. (2008) Multiplex caspase activity and cytotoxicity assays. In *Methods Mol. Biol.* **414**, 151–162. Apoptosis and Cancer, Methods and Protocols, ed. G. Mor and A. B. Alvero. Humana Press.
14. Riss, T. L. and Moravec, R. A. (2004) Use of multiple assay endpoints to investigate the effects on incubation time, dose of toxin, and plating density in cell-based cytotoxicity assays. *Assay Drug Dev. Technol.* **2**, 51–62.
15. Chatterjee, R. (2007) Cases of mistaken identity. Science **315**, 928–931.
16. Inglese, J., Auld D. S., Jadhav, A., Johnson, R. L., Simeonov, A., Yasgar, A., Zheng, W., and Austin, C. P. (2006) Quantitative high-throughput screening: A titration-based approach that efficiently identifies biological activities in large chemical libraries. *Proc. Natl. Acad. Sci. USA* **103**, 11473–11478.

Chapter 13

Fluorescein Diacetate for Determination of Cell Viability in 3D Fibroblast–Collagen–GAG Constructs

Heather M. Powell, Alexis D. Armour, and Steven T. Boyce

Abstract

Quantification of cell viability and distribution within engineered tissues currently relies on representative histology, phenotypic assays, and destructive assays of viability. To evaluate uniformity of cell density throughout 3D collagen scaffolds prior to in vivo use, a nondestructive, field assessment of cell viability is advantageous. Here, we describe a field measure of cell viability in lyophilized collagen–glycosaminoglycan (C–GAG) scaffolds in vitro using fluorescein diacetate (FdA). Fibroblast–C–GAG constructs are stained 1 day after cellular inoculation using 0.04 mg/ml FdA followed by exposure to 366 nm UV light. Construct fluorescence quantified using Metamorph image analysis is correlated with inoculation density, MTT values, and histology of corresponding biopsies. Construct fluorescence correlates significantly with inoculation density ($p < 0.001$) and MTT values ($p < 0.001$) of biopsies collected immediately after FdA staining. No toxicity is detected in the constructs, as measured by MTT assay before and after the FdA assay at different time points; normal in vitro histology is demonstrated for the FdA-exposed constructs. In conclusion, measurement of intracellular fluorescence with FdA allows for the early, comprehensive measurement of cellular distributions and viability in engineered tissue.

Key words: Quality assurance, Tissue engineering, 3D scaffold, Cell viability, Intracellular fluorescence

1. Introduction

Engineered tissues have the potential to replace or repair diseased or damaged tissues. Their outcome in vivo is highly correlated with their viability and function prior to implantation (1–3). A large number of engineered tissues are fabricated by the inoculation of single or multiple cell types onto three-dimensional (3D) tissue engineering scaffolds (4–7). For example, engineered skin is formed by serial inoculation of human autologous fibroblasts and keratinocytes on a 3D collagen scaffold (8–11). Engineered blood vessels can be made by inoculating vascular smooth muscle cells on

Martin J. Stoddart (ed.), *Mammalian Cell Viability: Methods and Protocols*, Methods in Molecular Biology, vol. 740,
DOI 10.1007/978-1-61779-108-6_13, © Springer Science+Business Media, LLC 2011

3D polyurethane scaffolds (12). Autologous endothelial cell-seeded heart valves are designed to minimize thrombogenicity postimplantation (13). Bone tissue has been engineered by seeding human mesenchymal stem cells on decellularized trabecular bone (14). As cell distribution, density, and viability are intimately linked with tissue function, it is essential to evaluate uniformity of cell distribution and viability within a 3D scaffold to assess the suitability of engineered tissues for clinical application/implantation. Quality assurance of engineered tissues is commonly determined using histology, functional assays of differentiated phenotypes, such as surface hydration or transepidermal water loss for skin (15–17), contractility in response to endothelin-1 for vessels (18), platelet activation and adherence assays for heart valves (13), and metabolic viability assays such as the 3-(4,5-dimethylthiazol-2-yl)-2,5-diphenyltetrazolium bromide (MTT) assay (11, 14, 17, 19). These assessments are often destructive, thereby reducing the tissue available for clinical use, and/or can only assay a small fraction of the total tissue area. Thus, a nondestructive assay of cell viability in the entire engineered tissue would represent an important advancement in preclinical quality assurance.

Fluorescein has been used in the laboratory to assess the in vitro viability of a wide variety of cell types and tissues including human fetal cerebral cortical cells (20), human keratinocytes (21), and ovine articular cartilage (22), and to assess skin flap viability (23) and burn depth (24) in vivo. By virtue of its bipolar side chains, fluorescein diacetate (FdA) easily penetrates the cell membrane. FdA remains colorless until the acetate moieties in FdA are cleaved nonspecifically, by intracellular esterases which convert the non-fluorescent FdA to fluorescein (25). Metabolically active cells with intact cell membranes can then be visualized under ultraviolet (UV) light (26, 27); the intensity of fluorescence seen under UV light is therefore directly proportional to the number of viable cells.

FdA staining and computer-assisted planimetry allow direct visualization of cell distribution and viability in vitro. The FdA assay accurately reflects the presence of viable cells when compared to traditional destructive assays and can be used to predict viability at a later time points with little to no cytotoxicity. Care must be taken when performing the assay to properly quantify scaffold autofluorescence and to calibrate the FdA intensity to known cell densities.

2. Materials

2.1. Cell Culture and Lysis

1. Dulbecco's Modified Eagle's Medium supplemented with 5% fetal bovine serum (FBS) and 1% antibiotic–antimycotic for dermal fibroblast culture.

2. HEPES-buffered saline (HBS): HEPES is dissolved in tissue culture water (see Note 1) at 30 mM, with 10 mM dextrose, 3 mM potassium chloride, 0.13 M sodium chloride, and 1 mM sodium phosphate heptahydrate, pH to 7.40, sterile filter, and stored at 4°C.
3. Solution of trypsin (0.025%) and ethylenediamine tetraacetic acid (EDTA, 1 mM) in HBS.
4. Water bath.
5. Fetal bovine serum.
6. Hemocytometer.
7. Inverted brightfield microscope (Nikon, Melville, NY).
8. Refrigerated tabletop centrifuge.
9. Sterile stereological and aspirating pipettes.

2.2. 3D Collagen–GAG Scaffolds

1. Fibrous collagen from bovine hide (Kensey Nash, Exton, PA) dissolved in 0.5 M acetic acid at 0.6 wt/vol%.
2. GAG–acetic acid solution: Chondroitin-6-sulfate dissolved in acetic acid at 0.35 wt/vol%.
3. Cold room (4°C) (see Note 2).
4. Digital stir plate and stir bar.
5. Syringe pump.
6. 60 ml syringe.
7. Casting frame: Rectangular metallic mold (aluminum plate and bar stock, Lyon Industries, South Elgin, IL) formed out of two sheets of metal separated by a 1-mm thick and 5-mm wide rubber spacer (Small Parts Inc., Miramar, FL) fitted around the edges of the metal. A cavity (20 cm × 10 cm × 1 mm) is created by the spacers.
8. –80°C Freezer.
9. 95% Ethanol (EtOH) bath. 2 l of ethanol poured into a 24 in. × 12 in. metal container.
10. Lyophilizer-Virtis Advantage XL (SP Industries, Gardiner, NY).
11. Vacuum oven.
12. Cryo-gloves.

2.3. Cell-Scaffold Construct Preparation

1. Cells from Subheading 2.1.
2. 150-mm cell culture dish.
3. Variable speed pipette-aid.
4. Merocel (Medtronic/XoMed, Minneapolis, MN).
5. N-terface (Winfield, TX).
6. Blunt edge forceps.

7. Bunsen burner.
8. Glass beaker containing 95% ethanol.

2.4. FdA Assay

1. FdA dissolved in acetone at 5 mg/ml and stored protected from light at 4°C.
2. UV lightbox (Fisher BioTech).
3. Camera hood (Electrophoresis Systems, Fisher).
4. Polaroid black and white film type 667.
5. Dell 1815 Scanner (see Note 3).
6. Image Analysis software (Metamorph) (see Note 4).
7. Snapwell cell culture inserts and tissue culture plate (Corning, NY).
8. Sterile surgical gloves.
9. Sterile 12 mm diameter biopsy punch (see Note 5).

2.5. Correlation of FdA Assay to MTT Assay and Total DNA Content

1. Thiazolyl blue tetrazolium bromide (MTT) dissolved in phosphate-buffered saline (pH 7.4) at 0.5 mg/ml, sterile filter, and stored at 4°C.
2. 2-Methoxy ethanol.
3. UV/Vis Spectrophotometer with 548-nm filter (Spectracount, Packard Bioscience Corporation, Meriden, CT).
4. Easy-DNA kit (Invitrogen).
5. ND-1000 spectrophotometer (Nanodrop, Wilmington, DE).
6. Statistical software (SigmaStat, San Jose, CA).

3. Methods

FdA rapidly penetrates the cellular plasma membrane and is cleaved to form fluorescein. The FdA solution at 0.04 mg/ml is sufficiently concentrated to allow detection of cells above any scaffold autofluorescence. Correlations between intensity of fluorescence can be made with cell number, punch biopsy histology, and viability. Assessment with FdA provides a global, nondestructive measure of cell distribution and viability within 3D scaffolds.

We developed a standardized, quantitative assay using FdA to evaluate fibroblast density and distribution across a collagen–glycosaminoglycan scaffold. This assay was necessary for us to be able to monitor cell inoculation uniformity when using larger surface area scaffolds or different scaffold preparations. The methods described here outline the steps to prepare a standard collagen scaffold inoculated with fibroblasts, and to evaluate its cell density 1 day later using the FdA assay. The fibroblast-inoculated scaffolds are immersed in the FdA solution at a nontoxic concentration, photographed immediately under fluorescent conditions, and rinsed prior to being returned to the incubator in fresh culture media. We also outline the techniques for and results of correlating the nondestructive FdA assay with the destructive MTT assay and DNA content measurements with this construct. The toxicity of the FdA assay has been evaluated with respect to keratinocytes, fibroblasts, fibroblast-inoculated scaffolds, and keratinocyte–fibroblast-inoculated scaffolds (28). Fluorescence can be quantified according to (1) the average over a defined surface area (Fig. 2), (2) the degree of variability between the maximum and minimum fluorescence over a given surface, or (3) the percent surface area above a defined threshold fluorescence. By establishing standard curves for fluorescence by inoculation density, MTT assay, or total DNA content, a suitable threshold can be determined for specific cell-scaffold combinations.

3.1. Preparation of Collagen Scaffolds

1. The raw collagen from bovine hide is mixed with 0.5 M acetic acid at 300 rpm on a stir plate within a refrigeration room (see Note 6). Mix continuously for 24 h.
2. Calculate the amount of GAG–acetic acid solution required for scaffold fabrication (add 1 ml of GAG solution for every 6.5 ml of collagen solution).
3. Fill the syringe with the appropriate amount of solution. Using the syringe pump, slowly add (~12 ml/h) GAG–acetic acid mixture to collagen–acetic acid solution. Mix for a total of 6 h.
4. Pour solution into a casting frame and freeze in the ethanol bath at −80°C for 1 h.
5. Remove the top section from the casting mold (see Note 7) and place the bottom section and frozen collagen–GAG solution into prechilled lyophilizer.
6. Lyophilize the frozen solvent and place the resultant porous collagen–GAG scaffold into vacuum oven set to 140°C and 30 mmHg for 24 h.
7. Cut collagen–GAG scaffold to desired size and sterilize by placing scaffold into a sterile culture dish filled with 70% EtOH. Procedure performed in class II biological safety cabinet (BSC). Alternatively, the scaffolds may be packaged into peel packs, and sterilized by gamma-irradiation (25 kGy) for later use.
8. Rinse with HBS for 10 min, repeat four times followed by 2–30 min rinses in cell culture medium (see Note 8).

3.2. Cell Harvesting

1. Warm fibroblast medium, HBS, and trypsin to 37°C in water bath.
2. Supplemented fibroblast culture medium is used to neutralize trypsin in fibroblast cultures.
3. Once fibroblasts have reached approximately 80% confluence, remove from incubator, aspirate medium, and rinse once with HBS for 2 min (see Note 9).
4. Aspirate HBS, add trypsin to each flask, and reclose flask cap. Place flasks in incubator for approximately 2 min.
5. Tap side of flask with hand to detach cells from growth surface.
6. Add equivalent amount of supplemented medium to flasks, rinse growth surface of flask with medium three times, remove cell-containing medium, and place into sterile conical tube.
7. Centrifuge at 1,000 rpm ($200 \times g$) for 7 min.
8. Resuspend and count cells using hemocytometer.

3.3. Preparation of Fibroblast-Scaffold Constructs

1. Using flame sterilized forceps, place a sterile piece of Merocel into a 150-mm culture dish. Immerse in cell culture medium. Rinse twice.
2. Place an 8 cm × 8 cm square of sterile N-terface on the Merocel followed by the collagen scaffold (see Note 10).
3. Aspirate excess medium from Merocel containing dish.
4. Inoculate collagen scaffold with desired density of fibroblasts (e.g., $1 \times 10^4/\text{cm}^2$).
5. Add medium to dish until medium level reaches the top of the Merocel. Do not allow medium to flow over the top of the Merocel. Incubate at 37°C, 5% CO_2 for 3 h (see Note 11).
6. Lift cell-scaffold constructs on the N-terface from the Merocel and place into a 150-mm cell culture dish containing 50 ml medium.
7. Culture for up to 21 days with medium changed daily.

3.4. FdA Assay

1. Warm HBS and cell culture medium to 37°C.
2. Wipe UV lightbox with 70% EtOH and place into BSC (see Note 12).
3. Using stock solution of FdA in acetone, mix a fresh dilution of 0.04 mg/ml of FdA in HBS. Sterile filter into foil covered bottle (see Note 13).
4. Remove cell–collagen construct from incubator and aspirate medium (see Note 14).
5. Add FdA solution being careful to fully saturate all sides of construct. Allow construct to soak in solution for 20 min at room temperature.

6. Aspirate FdA solution and rinse cell–collagen construct thoroughly with HBS.
7. Transfer cell–collagen construct to empty sterile 150-mm cell culture dish and place dish on UV lightbox (see Note 15).
8. Turn lightbox on for 20 s to convert the nonfluorescent FdA to fluorescein.
9. Wipe edges of camera hood with 70% EtOH.
10. Place camera hood over the dish to block out all other light. Turn on UV lightbox and take picture (1/8 s shutter speed) after lightbox has been on for 5 s (see Note 16).
11. Allow photo to develop for 60 s.
12. While photos are drying, transfer cell–collagen constructs back into cell culture dish containing medium and incubate. After 3 h, change medium to ensure that there is no residual FdA solution within the construct.
13. Scan photographs at 300 dpi in tiff format (Fig. 1).
14. Measure average fluorescence intensity using Metamorph software (see Note 17).
15. In Metamorph, use the select area tool to select just the cell–collagen construct within the photograph. Measure grayscale intensity values including average, high, and low. The average fluorescence intensity for the acellular grafts (0 cells/cm^2) or the background was subtracted from the cell containing groups (Fig. 1).

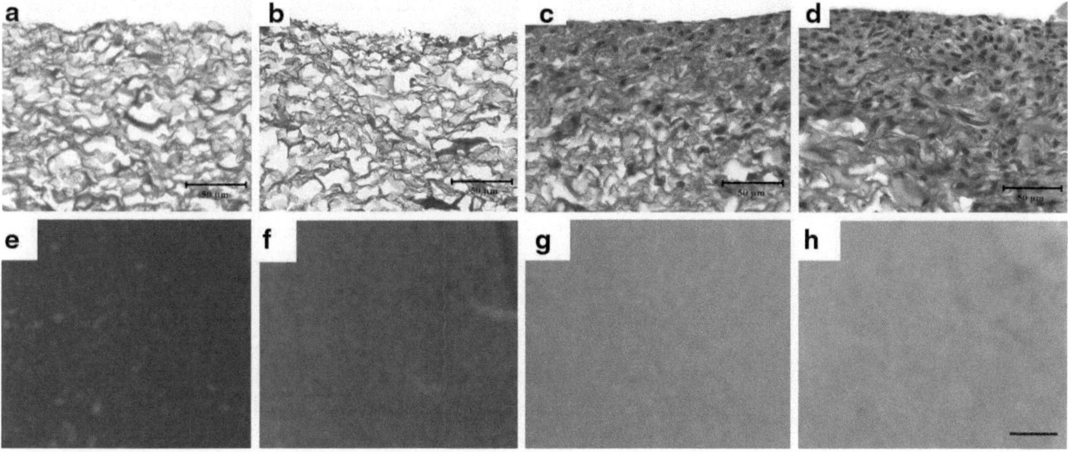

Fig. 1 Qualitative assessment of collagen–GAG scaffolds inoculated with fibroblasts at variable densities; 0 cells/cm^2 (**a**, **e**), 5E4 cells/cm^2 (**b**, **f**), 5E5 cells/cm^2 (**c**, **g**), and 5E6 cells/cm^2 (**d**, **h**). *Black* and *white* photographs of fluorescein intensity via exposure to FdA–HBS (5 mg/ml) for 20 min and H&E stained cross-sectional histology of collagen–GAG scaffolds embedded in paraffin (**a**, **e**). *Scale bar* = 200 μm (**a**–**d**) and 1 cm (**e**–**h**). (Reproduced from ref. 28, with permission from Mary Ann Liebert Inc.).

3.5. Standard Curves for FdA Intensity, Inoculation Density, DNA Content, and MTT Assay

1. Prepare collagen scaffolds as in Subheading 3.3.
2. Use sterile biopsy punch to cut 15 mm diameter circles in the rinsed collagen scaffold or use a sterile scalpel to cut small squares approximately 15 mm × 15 mm in size.
3. Place sterile Snapwell inserts and 6-well plate into BSC and remove lid.
4. Using sterile technique, remove lower ring from Snapwell insert (see Note 18).
5. Place lower ring with collagen scaffold back onto upper portion of insert affixing the collagen scaffold to the insert.
6. Harvest cells following Subheading 3.2. Inoculate cells onto collagen within Snapwell insert at 0, 5.0e3, 5.0e4, 5.0e5, and 5.0e6/cm^2. Adjust the total volume of inoculum to range between 250 and 500 μl.
7. Incubate for 1 day and perform FdA assay. Assay can be performed with scaffolds in the inserts or scaffolds can be removed prior to the assay.
8. After the assay has been performed, return Snapwell inserts to 6-well plate containing medium. Rinse twice with medium (30 min per rinse, return plate to incubator after each rinse).
9. Remove cell–collagen constructs from Snapwell insert and place into a new 6-well plate containing 4 ml of MTT–PBS solution.
10. Incubate for 3 h.
11. Remove MTT–PBS solution and add 4 ml of methoxy ethanol in a chemical fumehood. Place on a plate shaker for 3 h at room temperature.
12. Remove cell-scaffold constructs and read optical density on a spectrophotometer at 590 nm.
13. Perform steps 1–8 again with a new set of fibroblast-scaffold constructs.
14. After rinsing the constructs, remove the construct and perform the DNA quantification following the instructions provided by the Easy-DNA kit.
15. Use statistical software to correlate inoculation density, MTT absorbance values, and DNA content with FdA intensity values. For fibroblasts on collagen scaffolds the relationship is linear but this may not be true for all cell types on all types of scaffolds (Fig. 2) (see Note 19).
16. The standard curves can then be utilized to extract additional quantitative data from the FdA intensity values.

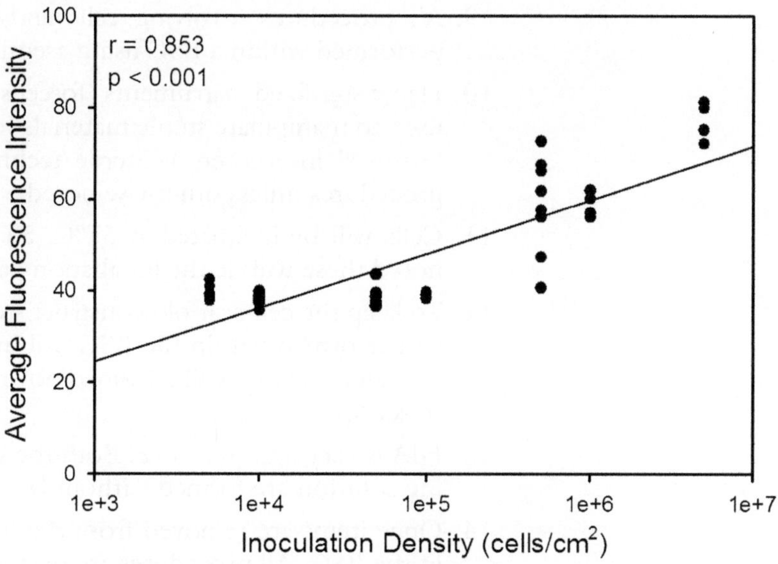

Fig. 2. Quantitative correlation between average fluorescein intensity measured with Metamorph software and fibroblast inoculation density 1 day after fibroblast inoculation. (Reproduced from ref. 28 with permission from Mary Ann Liebert Inc.).

4. Notes

1. All solutions should be prepared in water with a resistance of 18.2 MΩ cm. This is referred to as "water" in the text.
2. Cold room can be substituted by a stir plate within a 4°C refrigerator.
3. Most general office scanners will suffice. Caution is taken to avoid scanning the images at resolutions less than 300 dpi as significant pixilation can occur.
4. Other image software such as ImageJ may be used to calculate average grayscale intensity.
5. A biopsy punch is not essential. Collagen scaffolds can be cut into small square pieces using a sterile scalpel and prep blade or sterile scissors.
6. The quantity of collagen–GAG solution required is proportional to the area of collagen scaffold needed. 30 ml will form approximately 15 cm^3 ($15 \times 10 \times 0.1$ cm), or 150 cm^2.
7. When removing frozen collagen–GAG mold from casting frame, caution is required to open frames quickly and gently to avoid cracking.
8. Any commercially available scaffold can also be utilized for this assay. Be sure to quantify the autofluorescence for each scaffold type used.

9. All procedures involving cells and/or sterile scaffolds are performed within a BSC using aseptic technique.
10. Flame sterilized instruments (forceps, scalpel, prep blade) are used to manipulate sterile materials (collagen–GAG scaffolds, Snapwell inserts, etc.). Sterile technique is assumed for all procedures unless otherwise noted.
11. Cells will be incubated at 37°C, 5% CO_2. Unless otherwise noted these will be the incubation parameters.
12. To keep the cell-scaffold construct sterile, the entire assay will be performed within the BSC. All instruments brought into the safety cabinet will be wiped down with 70% ethanol prior to set up.
13. FdA is very light sensitive. Both the stock solution and working solution are formed without laboratory lights on.
14. Once items are removed from the incubator they are placed in the BSC. All procedures are performed in a BSC from this point forward.
15. Fluorescein can be excited at a wide range of wavelengths, with a maximal excitation at 494 nm (29). The FdA assay may be used either qualitatively or quantitatively. Following cell inoculation, cell adherence and localization on the matrix can be visualized topographically and evaluated for uniformity for quality control of the inoculation and collagen scaffold production methods. Such a qualitative analysis would require only a handheld Wood's lamp to visualize the fluorescence, as early as day 3 after inoculation. Alternatively, cell density and distribution on the matrix can also be quantified, in order to compare results between the groups.
16. Personal protective equipment (UV goggles or face shield) is required to avoid physical injury due to the UV light.
17. Quantification requires strict adherence to sterile cell culture techniques, including the use of a fluorescent UV light box and black and white camera in the laminar flow hood. The surface area which can be analyzed in this manner depends on the size of the camera hood available to capture fluorescent images. Once the photographic data is obtained, the fluorescence can be quantified using an image software program. The localization of the fluorescence can also be quantified in terms of the percent of the total surface area fluorescing over a defined threshold.
18. Sterile gloves can be donned to remove the lower rig from the Snapwell insert and for the insertion of the collagen–GAG scaffold.
19. The FdA assay can also be correlated to other biological properties such as total protein content.

Acknowledgments

HMP thanks the Shriners Hospitals for Children for a Post Doctoral Research Fellowship and grants 8507 and 8450 that supported these studies. ADA thanks the Fondation du CHUM for salary support. The authors also thank the NIH for financial support under grant GM50509 to STB. The authors thank Dr. George Babcock for the use of the Nanodrop spectrophotometer. The authors acknowledge Laura James for statistical analyses, Deanna Snider for tissue processing, and Jenny Klingenberg, Kevin McFarland and Todd Schuermann for technical assistance.

References

1. Stevens KR, Kreutziger KL, Dupras SK, Korte FS, Regnier M, Muskheli V, Nourse M B, Bendixen K, Reinecke H, Murry CE. (2009) Physiological function and transplantation of scaffold-free and vascularized human cardiac muscle tissue. *PNAS* **106**, 16568–73.
2. Capetti G, Ambrosio L, Savarino L, Granchi D, Cenni E, Baldini N, Pagani S, Guizzardi S, Causa F, Giunti A (2003) Osteoblast growth and function in porous poly epsilon-caprolactone matrices for bone repair: a preliminary study. *Biomaterials* **24**, 3815–24.
3. Park JC, Hwang YS, Suh H. (2000) Viability evaluation of engineered tissues. *Yonsei Medical Journal* **41**, 836–44.
4. Rezwan K, Chen QZ, Blaker JJ, Boccaccini AR. (2006) Biodegradable and bioactive porous polymer/inorganic composite scaffolds for bone tissue engineering. *Biomaterials* **27**, 3413–31.
5. Jones JR, Ehrenfried LM, Hench LL. (2003) Optimising bioactive glass scaffolds for bone tissue engineering. *Biomaterials* **27**, 964–73.
6. Hutmacher DW, Ng KW, Kaps C, Sittinger M, Klaring S (2003) Elastic cartilage engineering using novel scaffold architectures in combination with a biomimetic cell carrier. *Biomaterials* **24**, 4445–58.
7. Ushida T, Furukawa K, Toita K, Tateishi T. (2002) Three-dimensional seeding of chondrocytes encapsulated in collagen gel into PLLA scaffolds. *Cell Transplant* **11**,489–94.
8. Powell HM, Boyce ST. (2009) Engineered Human Skin Fabricated Using Electrospun Collagen-PCL Blends: Morphogenesis and Mechanical Properties. *Tissue Eng Part A* **15**, 2177–87.
9. Lee W, Debasitis JC, Lee VK, Lee JH, Fischer K, Edminster K, Park JK, Yoo SS. (2009) Multilayered culture of human skin fibroblasts and keratinocytes through three-dimensional freeform fabrication. *Biomaterials* **30**, 1587–95.
10. MacNeil S. (2008) Biomaterials for tissue engineering of skin. *Materials Today* **11**, 26–35.
11. Powell HM, Supp DM, Boyce ST. (2008) Influence of electrospun collagen on wound contraction of engineered skin substitutes. *Biomaterials* **29**, 834–43.
12. Dahl SLM, Rhim C, Song YC, Niklason LE. (2007) Mechanical properties and compositions of tissue engineered and native arteries. *Ann Biomed Eng* **35**, 348–55.
13. Kasimir MT, Rieder E, Seebacher G, Nigisch A, Dekan B, Wolner E, Weigel G, Simon P. (2006) Decellularization does not eliminate thrombogenicity and inflammatory stimulation in tissue-engineered porcine heart valves. *J Heart Valve Dis* **15**, 278–86.
14. Grayson WL, Bhumiratana S, Cannizzaro C, Chao PH, Lennon DP, Caplan AI, Vunjak-Novakovic G. (2008) Effects of initial seeding density and fluid perfusion rate on formation of tissue-engineered bone. *Tissue Eng Part A* **14**,1809–20.
15. Barai ND, Supp AP, Kasting GB, Visscher MO, Boyce ST. (2007) Improvement of epidermal barrier properties in cultured skin substitutes after grafting onto athymic mice. *Skin Pharmacol Physiol* **20**,21–28.
16. Supp AP, Wickett RR, Swope VB, Harriger MD, Hoath SB, Boyce ST. (1999) Incubation of cultured skin substitutes in reduced humidity promotes cornification in vitro and stable

engraftment in athymic mice. *Wound Repair Regen* 7, 226–37.

17. Powell HM, Boyce ST. (2008) Fiber density of electrospun gelatin scaffolds regulates morphogenesis of dermal-epidermal skin substitutes. *J Biomed Mater Res Part A* **84A**, 1078–86.
18. Laflamme K, Roberge CJ, Grenier G, Remy-Zolghadri M, Pouliot S, Baker K, Labbe R, D'Orleans-Juste P, Auger FA, Germain L. (2006) Adventitia contribution in vascular tone: insights from adventitia-derived cells in a tissue-engineered human blood vessel. *FASEB Journal* **20**, E516– 24.
19. Sun T, Mai SM, Norton D, Haycock JW, Ryan AJ, MacNeil S. (2006) Self-organization of skin cells in three-dimensional electrospun polystyrene scaffolds. *Tissue Eng* **11**, 1023–33.
20. Zhang F, Zhang P, Wang E, Su FZ, Qi YC, Bu X, Wang YF. (1995) Fluorescein diacetate uptake to determine the viability of human fetal cerebral cortical cells. *Biotech Histochem* **70**, 185–7.
21. Armeni T, Damiani E, Battino M, Greci L, Principato G. (2004) Lack of in vitro protection by a common sunscreen ingredient on UVA-induced cytotoxicity in keratinocytes. *Toxicology* **203**, 165–78.
22. Muldrew K, Hurtig M, Novak K, Schachar N, McGann LE. (1994) Localization of freezing injury in articular cartilage. *Cryobiology* **31**, 31–8.
23. Dvali LT, Dagum AB, Pang CY, Kerluke LD, Catton P, Pennock P, Mahoney JL. (2000) Effect of radiation on skin expansion and skin flap viability in pigs. *Plast Reconstr Surg* **106**, 624–9.
24. Gatti JE, LaRossa D, Silverman DG, Hartford CE. (1983) Evaluation of the burn wound with perfusion fluorometry. *J Trauma* **23**, 202–6.
25. Altman SA, Randers L, Rao G. (1993) Comparison of trypan blue dye exclusion and fluorometric assays for mammalian cell viability determinations. *Biotechnol Prog* **9**, 671–4.
26. Clarke JM, Gillings MR, Altavilla N, Beattie AJ. (2001) Potential problems with fluorescein diacetate assays of ell viability when testing natural products for antimicrobial activity. *J Microbiol Methods* **46**, 261–7.
27. Poole CA, Brookes NH, Clover GM. (1993) Keratocyte networks visualized in the living cornea using vital dyes. *J Cell Sci* **106**, 685–92.
28. Armour AD, Powell HM, Boyce ST. (2008) Fluorescein diacetate for determination of cell viability in tissue-engineered skin. *Tissue Eng Part C-Methods* **14**, 89–96.
29. Haughland RP. Handbook of Fluorescent Probes and Research Chemicals, 1996. Molecular Probes Inc, Eugene Oregon, 6th edition.

Chapter 14

Confocal Imaging Protocols for Live/Dead Staining in Three-Dimensional Carriers

Benjamin Gantenbein-Ritter, Christoph M. Sprecher, Samantha Chan, Svenja Illien-Jünger, and Sibylle Grad

Abstract

In tissue engineering, a variety of methods are commonly used to evaluate survival of cells inside tissues or three-dimensional (3D) carriers. Among these methods confocal laser scanning microscopy opened accessibility of 3D tissue using live cell imaging into the tissue or 3D scaffolds. However, although this technique is ideally applied to 3D tissue or scaffolds with thickness up to several millimetres, this application is surprisingly rare and scans are often done on slices with thickness <20 μm. Here, we present novel protocols for the staining of 3D tissue (e.g. intervertebral disc tissue) and scaffolds, such as fibrin gels or alginate beads.

Key words: Calcein AM, Ethidium homodimer-1, PKH26 tracker dye, Confocal laser scanning microscopy, Hydrogel, Tissue, Cell viability

1. Introduction

Cell viability (CV) is an important parameter in tissue engineering and culture studies to evaluate long-term survival of cells (1). Currently, there are a number of assays available to determine CV in tissue or three-dimensional (3D) scaffolds, including lactate dehydrogenase (LDH) staining (2, 3), calcein AM with ethidium homodimer-1 staining, e.g. Live/Dead® (CaAM/EthH), and cell counting after scaffold/tissue digestion (4, 5). The CaAM/EthH dyes can be used to stain living and dead cells directly in the tissue or scaffold. The CaAM is enzymatically hydrolysed into calcein in living cells, turning those fluorescent green. The EthH, on the other hand, is only able to enter cells with a compromised membrane and stains nucleic acid fluorescent red. It is also possible to count cells after tissue/scaffold digestion using different dyes or

Martin J. Stoddart (ed.), *Mammalian Cell Viability: Methods and Protocols*, Methods in Molecular Biology, vol. 740,
DOI 10.1007/978-1-61779-108-6_14,

stains which differentiate living from dead cells, such as CaAM or trypan blue (TB) (6). Live/Dead viability/cytotoxicity tests have the advantage that the live cell staining is not dependent on cell proliferation and is a non-radioactive assay unlike thymidine uptake and ^{51}Cr release assays (7–9). Stained cells can be observed under epifluorescence illumination and counted to determine the percentage of viable (or non-viable) cells. For cells grown in multiwell plates, an overall fluorescence per well can be determined using a fluorescence plate reader. For 3D specimens, confocal laser scanning microscopy (cLSM) has widely been applied. The primary advantage of the cLSM is its ability to produce optical sections through a 3D sample, e.g. a piece of tissue, that contain information from only one focal plane. By moving the focal plane of the instrument stepwise through the depth of the specimen, a series (stacks) of optical sections can be collected (10, 11). Because optical sectioning is non-invasive, the 3D distribution and relative spatial relationship of stained living as well as fixed cells can be observed with reasonable clarity. Another important feature is that the computer-controlled cLSM produces digital images which are amenable to image analysis and processing and can also be used to compute surface- or volume-rendered 3D reconstructions of the specimen. Accordingly, data from images can be processed and converted into cell counts and live/dead cell ratios, respectively.

It seems evident that cLSM requires a more costly setup than using simple cell counting with haemocytometer and TB exclusion staining or applying LDH with cryosectioning and counterstaining using propidium iodide. Table 1 summarises key features of cell viability methods used for tissue engineering and lists advantages and disadvantages.

Here, we provide hands-on protocols to set up the live/dead protocol for 3D tissue, i.e. Calcein AM/ethidium homodimer-1 (CaAM/EthH) staining and describe in detail three protocols for cell viability assessment: (1) staining protocol for tissue bits, such as cartilage or intervertebral disc tissue, (2) a protocol to track

Table 1
Advantages and inconveniences of three commonly used microscopic methods to determine cell viability

	Backup of the original carrier[a]	Cell morphology	Cell distribution	Equipment costs	Time consuming
LDH	++	+	++	+	–
CaAM/EthH	–	++	++	++	+
Trypan blue	–	–	–	–	++

LDH lactate dehydrogenase assay, *CaAM/EthH* calcein AM/ethidium homodimer-1, i.e. Live/Dead© assay
– Not appropriate, + useful, ++ very useful
[a]Histological sections can be kept for long-term usage

exogenous mesenchymal stem cells labelled with red tracker dye (PKH26) and endogenous cells within intervertebral disc tissue, and (3) a protocol to stain cells in fibrin carriers or alginate beads.

2. Materials

2.1. Stage for 3D Scanning

A stage for 3D scanning with good optical characteristics for inverted microscopy can be produced with the following materials.

1. Aluminium plate (dimensions 50 mm×80 mm×6 mm) (Fig. 1a, b).
2. Metal drill to cut out a circular hole (Ø 22 mm).

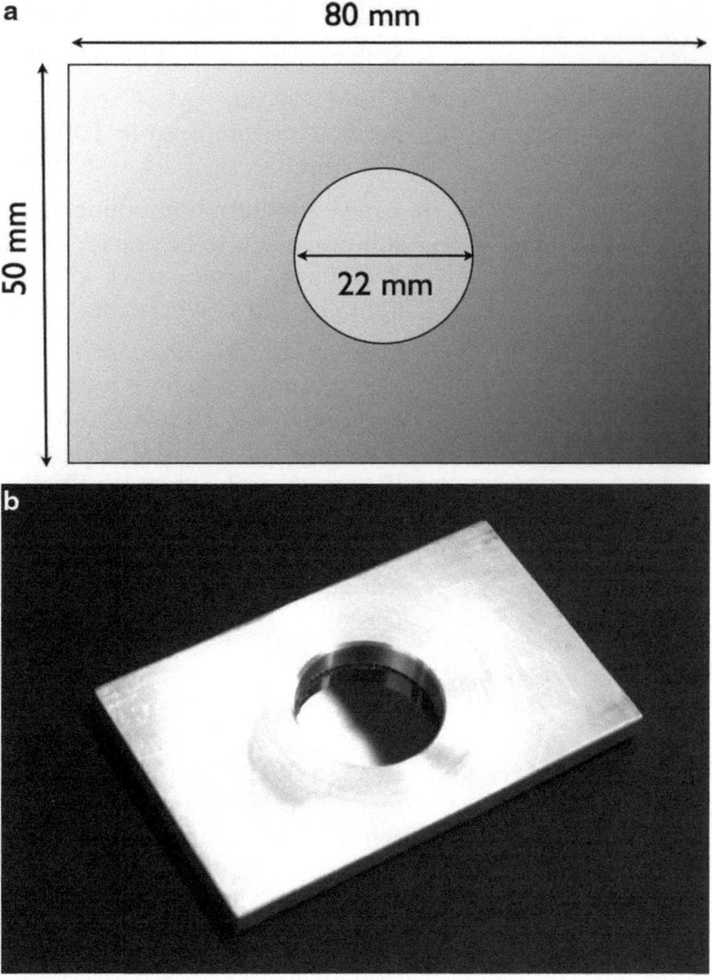

Fig. 1. Dimensions of simple custom-made sample holder (aluminium plate with 22 mm Ø drilled hole) used for improved 3D scanning. (**a**) schematic *top view* and (**b**) *side view*.

3. Coverslip 30 mm × 50 mm Nr.1 (e.g. Gerhard Menzel Glasbearbeitungswerk GmbH & Co. KG, Braunschweig, Germany).
4. Nusil© Medical Grade Silicon (MED-1137, Adhesive Silicone Type A, Silicone Technology, Carpinteria, CA). The Nusil© Medical Grade Silicon is dried for at least 72 h.
   The complete sample holder is washed twice with methanol and dried for 15 min before use.

2.2. Staining of Larger Tissues

1. Fresh intervertebral disc tissue, e.g. from bovine tails obtained from local slaughter house (ideally within hours post-mortem and skin removed).
2. Betadine™ solution.
3. 24-Well plate.
4. Phosphate-buffered saline (PBS).
5. Dulbecco's Modified Eagle's Medium (DMEM).
6. Scalpel blade holders #3 and #4.
7. Scalpel blades #10, #11, and #20.
8. 1 mM Calcein AM (CaAM) stock solution. Stock solution needs to be prepared in 100% dimethylsulphoxide (DMSO) and stored in the dark at <–20°C.
9. 1 mM ethidium homodimer-1 (EthHD) (see Note 1). Stock solution needs to be prepared in 80% DMSO and 20% ddH_20. Both fluorochrome stock solutions are stored protected from light at <–20°C where they are stable for at least 6 months.
10. Staining solution: Prepare 1 mL of staining solution per sample (~3 mm × 3 mm × 3 mm), using serum-free DMEM. For disc tissue, use 10 μM CaAM (1:100 dilution of stock solution) and 1 μM EthHD (1:1,000 dilution of stock solution). Prepare the dye "cocktail" directly before staining, since fluorescent dyes are unstable in contact with water. Keep in the dark if possible.
11. Aluminium sample holder (see Subheading 2.1).
12. Inverted confocal Laser Scanning Microscope Axiovert220m with motor-driven stage (LSM 510 series, Carl Zeiss, Jena, Germany).

2.3. PKH26 Cell Tracker Labelling

1. PKH26 Red Fluorescent Cell Linker Kit for general cell membrane labelling (PKH26GL-1KT). This kit is for general cell membrane labelling. It has been characterised in a number of model systems and has been found to be useful for in vitro cell labelling, in vitro proliferation studies, and long-term in vivo cell tracking.
2. Ethanol 100%.
3. Foetal calf serum (FCS).

4. High glucose DMEM supplied with antibiotics.
5. High glucose DMEM + 10% FCS.
6. Tyrode's balanced salt solution (TBSS).
7. Primary bovine bone marrow-derived mesenchymal stem cells, which are ≤Passage 4 (see Note 2).
8. 1 mM (1 mg/mL) CaAM stock solution. Stock solution needs to be prepared in 100% DMSO. Fluorochrome stock solutions are stored protected from light at <–20°C where they are stable for at least 6 months.
9. 1 mM 4′-6-diamidino-2-phenylindole (DAPI) stock. Stock solution is prepared by dissolving DAPI powder in distilled H_2O with vortex.
10. 10× Trypsin/EDTA mix solution. Dilute 1:10 with TBSS.
11. 6 mm biopsy punch.
12. Scalpel blade holder #3.
13. Scalpel blades #10 and #11.
14. Fresh intervertebral disc tissue from bovine tails and from local slaughter house (ideally within hours post-mortem).
15. Hamilton syringe #1710 (5–100 μL range) (Hamilton).
16. Sharp-headed 22-G needle for Hamilton syringe.
17. Centrifuge to spin down cells at 400–500 × *g*.
18. Aluminium sample holder (see Subheading 2.1).
19. Inverted confocal Laser Scanning Microscope Axiovert220m with motor-driven stage (LSM 510 series, Carl Zeiss, Jena, Germany).

2.4. Staining of 3D Fibrin Carriers

1. Bovine cartilage, for instance, from the femoral condyles of a 4-month-old calf obtained from the local abattoir, cut in small pieces (9–25 mm^3); or approximately 10×10^6 primary chondrocytes.
2. Tyrode's balanced salt solution (TBSS).
3. Pronase I (Roche).
4. Collagenase II (Worthington).
5. High glucose DMEM supplied with antibiotics.
6. Foetal calf serum (FCS).
7. Dimethylsulphoxide (DMSO).
8. Fibrinogen solution (Tisseel©, Baxter, Vienna, Austria), 50 mg/mL.
9. Fibrinogen dilution buffer (Baxter, Vienna, Austria).
10. Thrombin solution (Tisseel©, Baxter), 10 U/mL (see Note 3).
11. Thrombin dilution buffer (Baxter, Vienna, Austria).

12. Sterile mould to prepare fibrin gels (e.g. cylinder of dimensions Ø 8 mm × 4 mm height, thus ~200 mm^3 volume).
13. Sterile punch that fits precisely the diameter of the mould to remove the fibrin carrier.
14. 1 mM (1 mg/mL) CaAM stock solution (see Subheading 2.2, item 8).
15. 1 mM EthHD stock solution (see Subheading 2.2, item 9).
16. 24-Well plate.
17. Scalpel blade holder #3.
18. Scalpel blades #10 and #11.
19. Aluminium sample holder (see Subheading 2.1).
20. Inverted confocal Laser Scanning Microscope Axiovert220m with motor-driven stage (LSM 510 series, Carl Zeiss, Jena, Germany).

3. Methods

3.1. Stage for 3D Scanning

In fluorescent microscopy, it is important to work with the best possible optics. Therefore, a custom-made sample holder is manufactured. A hole of 22 mm diameter is drilled into the 6-mm thick aluminium plate. Using adhesive silicone, a coverslip is glued on one side of the hole (Fig. 1). The sample holder prevents the sample (e.g. disc) from drying out during the time at the cLSM. The objectives are corrected for the thickness of the coverslip used. It is important to make sure that the sample is not floating in the liquid (e.g. PBS or TBSS). Furthermore, dirt particles or air bubbles between the coverslip and the sample will change the optical result.

3.2. Staining of Larger Tissues (e.g. Intervertebral Disc Tissue)

Here, we describe the staining of larger tissues, e.g. an intervertebral disc sample of 3 mm × 3 mm × 3 mm.

3.2.1. Preparation of Tissue Samples

1. Freshly prepare intervertebral discs (IVD) from bovine tails obtained from the local slaughter house under a laminar flow or some protected area to prevent contamination. Disinfect the tail with Betadine™ solution prior to dissection.
2. During dissection of the discs, frequently sprinkle the tail with PBS to prevent dehydration.
3. Use a scalpel holder #4 with blade #20 to remove as much of the soft tissue as possible from the caudal spine to ensure easy localization of the IVD. Change blades frequently.
4. If necessary, remove the spinous and transverse processes of the vertebrae by using bone removal pliers (see Note 4).

5. After removing the surrounding tissue, pass on to blunt dissection to avoid cutting into the disc. Therefore, pull the blade in a ~30° angle along the vertebrae and the IVD to remove muscles, nerves, and blood vessels from the outer annulus and the vertebrae.
6. Dissect the disc with a scalpel holder #3 blade #10 by cutting through the transition zone between disc and endplate.
7. Separate the disc into nucleus pulposus (NP) and annulus fibrosus (AF) tissue with a 6-mm diameter biopsy punch (see Note 5).
8. The following cuts are done with a scalpel holder #3 blade #10 (each cut with a fresh blade). Cut the punched-out NP sagittally into halves. Divide the donut-shaped AF in quarters (triangles) and further dissected it into inner and outer AF.
9. Distribute 1 mL staining solution per well into a 24-well plate, cover the plate with aluminium foil, and keep it at 4°C until use (use the same day).
10. Immerse the samples in 1 mL of the freshly prepared staining solution in a 24-well plate (see Subheading 3.2.1) and incubate for 2 h at 37°C in 90% humidity and 5% CO_2 (see Note 6).
11. After incubation, transfer the tissue into a fresh well plate, wash 1× with PBS and immediately observe it by cLSM.

3.2.2. Scanning Procedure with the cLSM

1. For this procedure the Argon and the HeNe1 laser on the LSM 510 (Carl Zeiss, Göttingen, Germany) mounted on an Axiovert200m equipped with the Software LSM510 (Version 3.2) are used.
2. To visualise the green fluorochromes, the 488-nm wave length of the Argon laser, a main beam splitter 488/543, a beam splitter 545 nm, and a band pass 505–530 nm are used on the first channel.
3. The 543-nm wave length of HeNe1 laser, a main beam splitter 488/543, a beam splitter 545 nm, and a long pass 585 nm are used on the second channel to image the red fluorochromes.
4. Stacks are taken with a 10× objective Plan Neofluar 10×/0.3 (Carl Zeiss, Göttingen, Germany) at a 512×512 pixels (field size of 921.4 μm×921.4 μm) with the pinhole at 1 Airy unit, 50% image overlap, and 5.68 μm intervals. The stack thickness is between 200 and 300 μm.
5. The tissue is removed from the well plate immediately before live-cell scanning and placed onto the custom-made specimen holder (Subheadings 2.1 and 3.1).
6. Raw data are taken in .lsm file format, which can be used as input for Carl Zeiss Original software or in software of other producers (see Fig. 2).

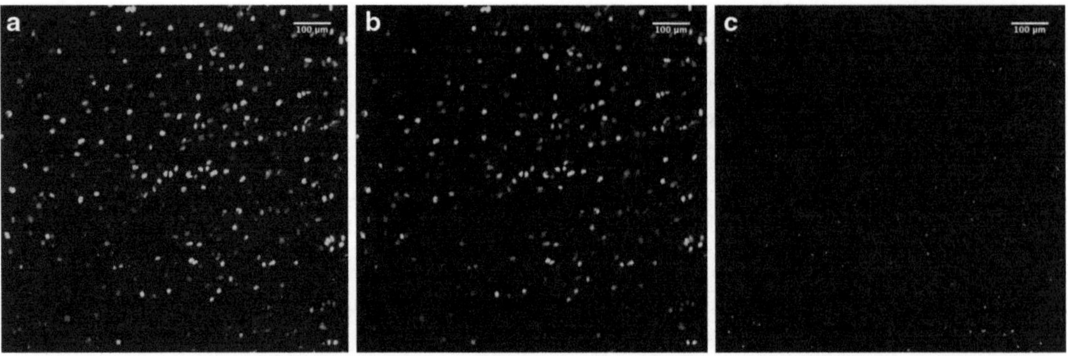

Fig. 2 (**a**) Representative scan of live/dead staining taken at 10× magnification through the nucleus pulposus of ovine caudal intervertebral disc, (**b**) *green* channel, and (**c**) *red* channel.

3.3. PKH26-Labelled Cell Tracking

Staining and injecting mesenchymal stem cells into whole bovine caudal IVD and tracking the cells with cLSM.

3.3.1. Staining of Mesenchymal Stem Cells with PKH26 Tracker Dye

1. Wash cells with TBSS.
2. Trypsinize bovine mesenchymal stem cells (bMSCs) (between Passages 2 and 4) at 80% confluency.
3. Add 20 mL 1× trypsin/EDTA solution to each T-300 flask and incubate at 37°C for 10 min.
4. Stop trypsin reaction by adding 20 mL DMEM with 10% FCS.
5. Transfer cell suspension into 50-mL Falcon® tube. Count number of cells and make sure the total number of cells for staining is less than 1×10^7.
6. Transfer the desired amount of cells into new 15-mL Falcon® tube.
7. Spin down cell suspension at $400 \times g$ for 5 min.
8. Wash cells with high glucose DMEM without serum and centrifuge at $400 \times g$ for 5 min.
9. Resuspend cells in 500 µL DILUENT C (provided by the kit).
10. Prepare working dye solution by adding 490 µL DILUENT C + 3 µL EtOH + 2 µL dye making up a final dye concentration of 4 µM.
11. Mix the prepared 495 µL working dye solution with the 500 µL cell suspension. Gently mix by inverting the tube two to three times and turn the 15-mL Falcon® tube up side down and incubate for 4 min at room temperature.
12. Stop the reaction by adding 1 mL FCS and incubate for 1 min.
13. Wash cells with 2 mL of DMEM with 10% FCS and centrifuge at $400 \times g$ for 5 min.

14. Transfer cells to new tubes and wash three times with TBSS.
15. Plate 2×10^6 PKH26 stained cells in T-150 flask in low glucose DMEM and let them recover overnight before injection.

3.3.2. Injection of bMSCs into the Intervertebral Disc

1. Wash labelled cells in T-150 flask with TBSS.
2. Trypsinize PKH26 labelled cells using 15 mL of 1× trypsin/EDTA solution.
3. Stop trypsin reaction by adding 20 mL of DMEM with 10% FCS.
4. Count the number of cells; prepare an aliquot for injection at 1×10^6 cells and centrifuge at $400 \times g$ for 5 min.
5. Resuspend the washed cells in 100 μL TBSS for injection.
6. Inject 50 μL of cell suspension into the centre of the IVD using 22-G 45°-sharp headed needle with the 100-μL Hamilton syringe. The injection speed should be approximately 50 μL/min.
7. The injected cells may be cultured for a certain time or imaged immediately on the confocal laser scanning microscope as described below.

3.3.3. Preparation of Tissue Samples and Staining Solutions

1. Remove endplates on both sides of the disc. Punch out nucleus pulposus (NP) using a 6-mm biopsy punch and make three more cuts with #11 scalpel blade in a triangle.
2. Dissect discs further into inner annulus fibrosus (AF) and outer AF. Cut inner and outer AF into four pieces.
3. Immerse each intervertebral disc tissue in 1 mL of TBSS containing 10 μL/mL CaAM stock (final concentration 10 μM) in a 24-well plate.
4. To ensure full penetration of the dye into the tissue, incubate tissue with staining solution at 4°C for 3 h and then 1 h at 37°C (see Note 6).
5. Take out the tissues and then incubate each of the tissue in 1 mL TBSS containing 2.5 μL DAPI stock solution at 37°C for 30 min.
6. Transfer tissues and store in TBSS prior imaging.

3.3.4. Scanning Procedure with the cLSM

1. For this procedure the Argon, the HeNe1, and the UV Laser (351 and 364 nm) on the LSM 510 (Carl Zeiss, Göttingen, Germany) mounted on an Axiovert200m equipped with the Software LSM510 (Version 3.2) are used.
2. The 351 and 364-nm wave lengths of UV laser are used on the first channel to scan the blue fluorochromes. The optical settings are main beam splitter UV/488/543/633 nm, the

first beam splitter 545 nm, the second beam splitter 490 nm, and a band pass 385–470 nm.

3. To visualise the green fluorochromes, the 488-nm wave length of the Argon laser, a main beam splitter UV/488/543/633 nm, a beam splitter 545 nm, and a band pass 505–530 nm are used.

4. The 543-nm wave length of HeNe1 laser, a main beam splitter UV/488/543/633 nm, a beam splitter 545 nm, and a long pass 560 nm are used to image the red fluorochromes.

5. Stacks are taken with a 10× objective Plan Neofluar 10×/0.3 (Carl Zeiss, Göttingen, Germany) at a 512 × 512 pixels (field size of 921.4 μm × 921.4 μm) and 50% image overlap and 4.78 μm intervals. The stack thickness is between 100 and 150 μm (Fig. 3).

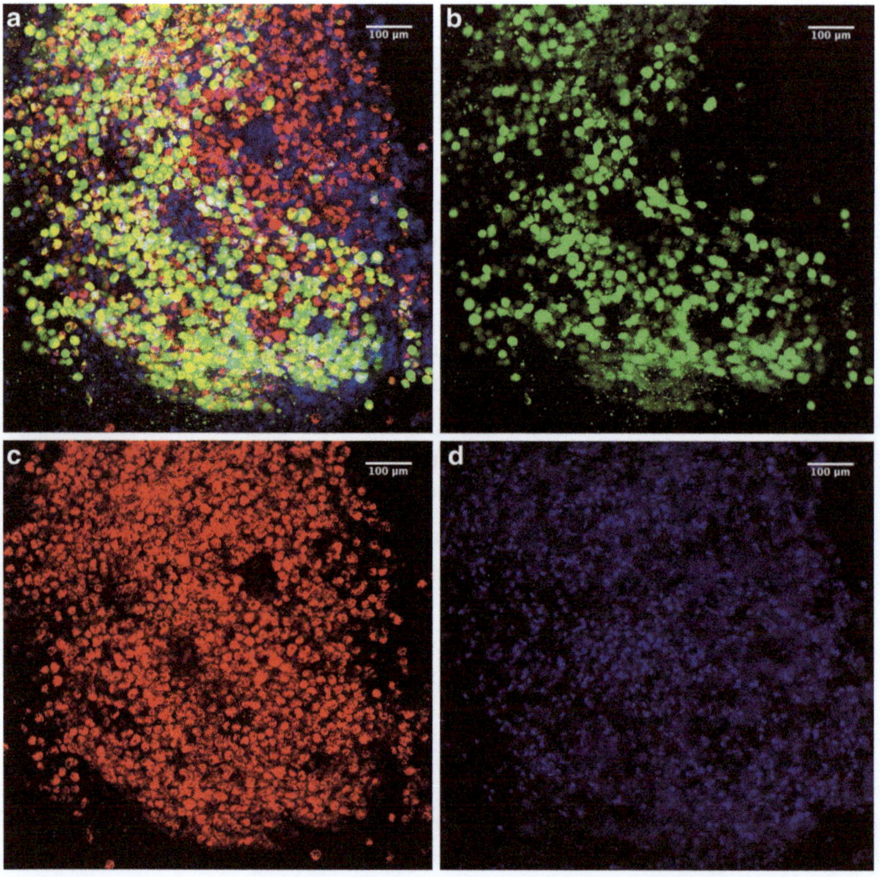

Fig. 3. (**a**) Representative scan of PKH26-labelled bovine mesenchymal stem cells (bMSCs) that were injected into a bovine intervertebral disc and stained with calcein AM (*green*) and DAPI at day 0 after injection. (**a**) Composite image – *green* and *blue* labelled cells mark living endogenous tissue and *blue* only denote dead disc cells. *Red* and *green* labelled cells denote living exogenous cells and *red* only dead exogenous cells, (**b**) *green* channel, (**c**) *red* channel, and (**d**) *blue* channel.

6. The tissue is removed from the well-plate immediately before live-cell scanning and placed onto the custom-made specimen holder (Subheadings 2.1 and 3.1).
7. Raw data are taken in ".lsm" file format, which can be used as input for Carl Zeiss Original software or in software of other producers.

3.4. Staining of 3D Fibrin Carriers

3.4.1. Protocol to Isolate and Expand Primary Bovine Chondrocytes

Bovine cartilage tissue pieces are washed with TBSS and pre-digested in spinner flasks with 1 mg/mL pronase I in TBSS for 2 h at standard conditions (37°C, 5% CO_2, 95% humidity). Pre-digested cartilage is then washed with TBSS and digested with 600 IU/mL collagenase II in DMEM by stirring overnight (14 h). After washing in TBSS, the primary chondrocytes are expanded up to P2 in DMEM, 10% FCS, 1% antibiotic/antimycotic solution and then frozen in DMEM containing 10% FCS and 10% DMSO. Cells are thawed and expanded up to P8 in DMEM, 10% FCS, 1% antibiotic/antimycotic solution in standard conditions for experimental use.

3.4.2 Preparation of Fibrin Samples and Staining Solutions

Fibrin carriers are polymerised in custom-made moulds of stainless steel (8 mm diameter × 4 mm height; final volume of 201.05 mm^3).

1. Fibrinogen and thrombin are prepared as intermediate solutions as follows: Tisseel© fibrinogen original recipe is diluted 1:1 with fibrinogen dilution buffer (recipe can be obtained from Baxter, Inc., Austria), resulting in a concentration of 50 mg/mL. Thrombin original concentration (500 U/mL) is diluted 50 times with thrombin dilution buffer.
2. Primary chondrocytes are trypsinized from culture flasks.
3. Cells are washed in PBS and suspended in fibrinogen solution.
4. Within 5 min, 100 μL of thrombin solution (10 U/mL) is added to 900 μL fibrinogen/cell suspension.
5. Immediately, 230 μL of this solution are transferred into the custom-made mould and left for 20 min at RT. The final cell density in the fibrin carriers is ~7,000 cells/mm^3, approximately corresponding to that observed in IVD and articular cartilage.
6. The fibrin carriers can be either analysed directly by proceeding to the staining section (see below) or cultured for some time (see Note 7).
7. Remove the fibrin carriers from the culture media.
8. Cut fibrin carriers sagittally into halves with a scalpel blade #11.

9. Immerse each half of fibrin carrier in 1 mL serum-free DMEM containing 10 µL/mL CaAM stock (final concentration 10 µM) and 1 µM EthHD in a 24-well plate.
10. To ensure diffusion of dyes throughout the sample incubate it with dyeing solution for 3 h at 4°C followed by 1 h in at 37°C, 5% CO_2, and 100% humidity.
11. Transfer and store samples in TBSS prior to imaging and image immediately.

3.4.3. Imaging Procedure

1. For this procedure, the Argon and the HeNe1 laser on the LSM 510 (Carl Zeiss, Göttingen, Germany) mounted on an Axiovert200m equipped with the Software LSM510 (Version 3.2) are used.
2. To visualise the green fluorochromes, the 488-nm wave length of the Argon laser, a main beam splitter 488/543, a beam splitter 545 nm, and a band pass 505–530 nm on the first channel are used.
3. The 543-nm wave length of HeNe1 laser, a main beam splitter 488/543, a beam splitter 545 nm, and a long pass 585 nm are used on the second channel to image the red fluorochromes.
4. Stacks are taken with a 10× objective Plan Neofluar 10×/0.3 (Carl Zeiss, Göttingen, Germany) at a 512×512 pixels (field size of 921.4 µm×921.4 µm) with the pinhole at 1 Airy unit, 50% image overlap, and 5.70 µm intervals.
5. Scan the carriers from top and bottom surfaces to ~200 µm depth at two random locations per side.
6. Count red and green cells (see Note 8).

4. Notes

1. Instead of the raw dyes for live-cell imaging, the dyes from Molecular Probes™ (Molecular Probes) may also be used. However, for certain applications in 3D tissue, we found it convenient to modify the stock concentration of the CaAM/EthHD. For IVD, we sometimes used a stock solution which was 5× higher, thus 5 mM.
2. Other primary cell types or cell lines may be used for injection.
3. The working concentrations of fibrinogen and thrombin can be achieved by dilution of the original Tisseel© solution (500 U/mL of thrombin and 100 mg/mL of fibrinogen) with specially provided buffers from Baxter. We use a fibrinogen

concentration of 50 mg/mL, which corresponded a 1:1 dilution, and a thrombin concentration of 10 U/mL (1/50 dilution of the original thrombin solution).

4. Depending on the level of the IVD or the species, the spinous processes of the vertebrae may obstruct the access to the transition zone between the disc and the cartilaginous end plates. In bovine, this applies only for the more proximal vertebrae, whereas in ovine, these processes are present throughout the entire caudal vertebrae.
5. Depending on the disc size, biopsy punches of different diameters are recommended: 6 mm punch: discs with a diameter ~15–20 mm; 5 mm punch: discs with a diameter ~9–14 mm.
6. The incubation time may vary with different type and size of the tissue used for staining. This should be optimised in preliminary experiments when using different tissues.
7. For the successful long-term culture of fibrin, addition of aprotinin or caproic acid may be necessary to prevent degradation of fibrin carrier through enzymatic activity (12, 13).
8. Red and green cells can be quantified per single image using a custom-made macro in ImageJ software (deposited at NIH, http://rsb.info.nih.gov/ij/plugins/index.html). The macro consists of a threshold step that passes a binary image with pixels in the range of 100–255 to the plug-in "nucleus counter" (McMaster University, Biophotonics Facility, http://www.macbiophotonics.ca/imagej), which then uses the "Otsu" method for particle counting. The minimum and maximum island sizes are set to 7–50 and 15–100 pixels for the red and green channels, respectively. The CV is then estimated on a subset of ten consecutive images, starting at the image with 50% or more of the maximum amount of the total cells per image in a single stack.

References

1. Gantenbein-Ritter, B., Potier, E., Zeiter, S., van der Werf, M., Sprecher, C. M., and Ito, K. (2008). Accuracy of three techniques to determine cell viability in 3D tissues or scaffolds. *Tissue Eng Part C, Methods* **14**, 353–8.
2. Stoddart, M. J., Furlong, P. I., Simpson, A., Davies, C. M., and Richards, R. G. (2006). A comparison of non-radioactive methods for assessing viability in ex vivo cultured cancellous bone: technical note. *Eur Cell Mater* **12**, 16–25; discussion 16–25.
3. Rauch, B., Edwards, R. B., Lu, Y., Hao, Z., Muir, P., and Markel, M. D. (2006). Comparison of techniques for determination of chondrocyte viability after thermal injury. *Am J Vet Res* **67**, 1280–5.
4. Catelas, I., Sese, N., Wu, B. M., Dunn, J. C., Helgerson, S., and Tawil, B. (2006). Human Mesenchymal Stem Cell Proliferation and Osteogenic Differentiation in Fibrin Gels in Vitro. *Tissue Eng* **12**, 1–12.
5. Haschtmann, D., Stoyanov, J. V., Ettinger, L., Nolte, L. P., and Ferguson, S. J. (2006). Establishment of a novel intervertebral disc/endplate culture model: analysis of an ex vivo in vitro whole-organ rabbit culture system. *Spine* **31**, 2918–25.

6. Melamed, M. R., Kamentsky, L. A., and Boyse, E. A. (1969). Cytotoxic test automation: a live-dead cell differential counter. *Science* **163**, 285–6.
7. Supino, R. (1995). MTT assays. *Methods Mol Biol* **43**, 137–49.
8. Mosmann, T. (1983). Rapid colorimetric assay for cellular growth and survival: Application to proliferation and cytotoxicity assays. *Journal of Immunological Methods* **65**, 55–63.
9. Oral, H. B., George, A. J., and Haskard, D. O. (1998). A sensitive fluorometric assay for determining hydrogen peroxide-mediated sublethal and lethal endothelial cell injury. *Endothelium* **6**, 143–51.
10. Lichtman, J. W. (1994). Confocal microscopy. *Scientific American* **271**, 40–5.
11. Boyde, A. (1994). Bibliography on confocal microscopy and its applications. *Scanning* **16**, 33–56.
12. Lee, C. R., Grad, S., Gorna, K., Gogolewski, S., Goessl, A., and Alini, M. (2005). Fibrin-polyurethane composites for articular cartilage tissue engineering: a preliminary analysis. *Tissue Eng* **11**, 1562–73.
13. Kupcsik, L., Alini, M., and Stoddart, M. J. (2009). Epsilon-aminocaproic acid is a useful fibrin degradation inhibitor for cartilage tissue engineering. *Tissue Eng Part A* **15**, 2309–13.

Chapter 15

Viability Assessment of Osteocytes Using Histological Lactate Dehydrogenase Activity Staining on Human Cancellous Bone Sections

Katharina Jähn and Martin J. Stoddart

Abstract

The assessment of viable osteocytes within bone tissue is of crucial importance. Osteocytes are the most abundant cells in bone. Due to their interconnectivity in the bone matrix they are hypothesised to play an important role in the maintenance of the extracellular matrix of bone. The death of osteocytes and the resulting disturbance of the osteocyte-canalicular network are responsible for the "loss of function" seen in several bone diseases. The lactate dehydrogenase (LDH) assay is a popular method to detect cell viability in bone sections. The major advantage of the LDH assay is the stability of the LDH enzyme for up to 36 h after cell death, eliminating any false negative viability results due to processing of the tissue. Here, we present a quick, reliable, and easy modification of the LDH assay using non-decalcified, thick, unfixed cancellous bone sections for the quantification of osteocyte viability.

Key words: Lactate dehydrogenase, Viability, Cancellous bone, Osteocyte

1. Introduction

The use of LDH activity staining as measurement for osteocyte viability goes back to 1982 when Wong et al. introduced this histological method for the use on bone sections derived from human femoral heads (1). He demonstrated that viable osteocytes were stained positive for LDH activity, while it was already known that necrotic tissue does not show LDH activity staining. The clear advantage of the LDH viability staining for the use on bone tissue is determined by the stability of the enzyme activity. LDH activity can be detected up to 36 h after cell death. This characteristic property of LDH enables the processing of mineralised bone tissue by cutting, drilling, or even freezing prior to viability staining without the creation of false-negative results.

Martin J. Stoddart (ed.), *Mammalian Cell Viability: Methods and Protocols*, Methods in Molecular Biology, vol. 740,
DOI 10.1007/978-1-61779-108-6_15,

This property of the LDH enzyme is of crucial importance, as bone possesses several characteristics that complicate viability analysis. Firstly, the assessment of cell viability within 3D samples of bone tissue is dependent on the reagent penetration. However, reagent penetration within bone is rather difficult. Cortical and even cancellous bone is considered as dense material due to the mineralisation of the extracellular matrix and the fine lacunar-canalicular-network. Secondly, additional misinterpretations of viability assessments within a 3D sample may arise due to the predominant uptake of a metabolically active reagent by surface cells, limiting the availability of reagent for the central cell population. Both effects are responsible for the creation of false-negative results within the centre of bone explants.

Lactate dehydrogenase (LDH, EC 1.1.1.27) is an ubiquitous cytoplasmic enzyme. The enzyme can be found in almost all living cells in a huge variety of organisms. The LDH enzyme is a tetrameric protein, which consists of polypeptide subunits of either type M or type H. Subunit M (M from muscle) is thought to be responsible for the reaction of the glycolysis product pyruvate and reduced nicotinamide adenine dinucleotide (NADH) to lactate and oxidised nicotinamide adenine dinucleotide (NAD^+) under the absence of oxygen, whereas the subunit H (H for heart) is more likely responsible for the reverse reaction (2) (Fig. 1).

The two polypeptide subunits M and H form in total five different LDH isoenzymes, LDH-1 (four H-subunits) is predominantly found in heart tissue, LDH-2 (three H- and one M-subunit) in the reticuloendothelial system, LDH-3 (two H- and two M-subunits) is found in lungs, LDH-4 (one H- and three M-subunits) in the kidneys, and LDH-5 (four M-subunits) is present in liver tissue as well as striated muscles. In humans, the polypeptide subunit M is encoded by LDHA (11p15.4) and subunit H by LDHB (12p12.2-p12.1). A third isoform known as LDHC is thought to have arisen by duplication of the LDHA sequence.

The classic example of the reversible LDH-driven reaction is seen during anaerobic respiration in muscle cells, when ATP is needed in large amounts while oxygen supply is restricted. Under these conditions LDH converts pyruvate and NADH to produce lactate and NAD^+. However, the reaction products are in equilibrium, so that the reverse reaction is also catalysed. The possible

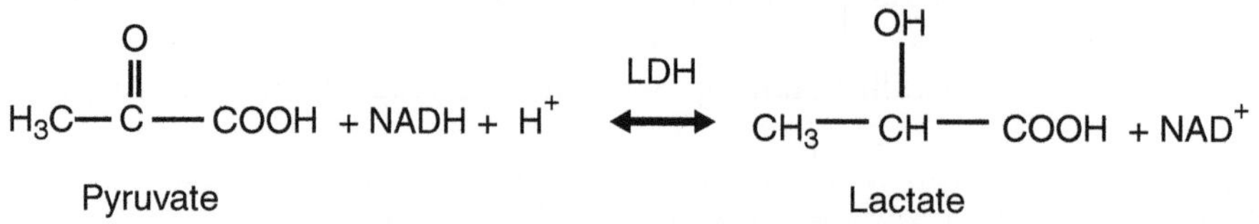

$$\underset{\text{Pyruvate}}{H_3C{-}\overset{\overset{\displaystyle O}{\|}}{C}{-}COOH} + NADH + H^+ \xleftrightarrow{\text{LDH}} \underset{\text{Lactate}}{CH_3{-}\overset{\overset{\displaystyle OH}{|}}{CH}{-}COOH} + NAD^+$$

Fig. 1. Schematic illustration of the reversible reaction catalysed by the LDH enzyme.

reaction mechanism involves the transfer of a hydride ion from C4 of NADH to the C2 of pyruvate and at the same time one proton is transferred from C2 of pyruvate to histidine 195 of the enzyme. Histidine 195 and arginine 171 stabilise the position of the substrate pyruvate/lactate by electrostatic bonding. Lactate is an end product of anaerobic respiration and can either be transported outside of the cell or directly reconverted to pyruvate. The majority of the produced lactate is transported to the liver, where during the cori-cycle, lactate is reconverted to pyruvate due to the action of the LDH enzyme. The produced pyruvate can then be converted to glucose during gluconeogenesis using adenosine triphosphate (ATP) of the liver. The so-produced glucose is transported back to the muscle where it is either stored in the form of glycogen or is directly used to produce muscle ATP (3).

The LDH assay introduced within this chapter is using the tetrazolium salt nitroblue tetrazolium (NBT) as third substrate next to lactate and NAD^+. The use of this particular substrate offers further advantages of the method. Bone matrix, especially in a highly mineralised state, is extremely autofluorescent complicating the use of fluorescent viability dyes, such as fluorescein diacetate or ethidium homodimer. During the LDH assay, the supplied tetrazolium salt is converted by viable cells to the water-insoluble dark-purple formazan. Due to the location of the LDH enzyme within the cytoplasm, one cell can be easily distinguished from another. Stoddart et al. demonstrated the use of the naturally occurring autofluorescence of bone matrix as background emission increasing the contrast to the positively dark-stained osteocytes that efficiently block the autofluorescence (4). Fluorescence micrographs are therefore taken at 450–490 nm excitation and 515–565 nm emission using a beam splitter at 510 nm. Central regions of the bone sections were visualised with these settings with a 20× objective lens. Due to the achieved depth of field of 4.12–4.52 μm, dark violet-stained osteocytes in focus could be accounted as viable osteocytes per bone area.

2. Materials

2.1. Bone Explant Preparation

1. 0.9% Sodium chloride solution pre-cooled to 4°C.
2. Hank's balanced salt solution (HBSS).
3. Exakt 300 band saw.
4. 10 mm diameter diamond tipped core bit (Synthes).
5. Annular saw (Leica).

2.2. LDH Staining

1. Polypep base solution: 5% Polypep, 2 mM Gly–Gly and 0.75% NaCl, pH 8.0 (stored at 4°C).
2. LDH staining solution: Add 60 mM lactic acid, 17.5 mg β-nicotinamide adenine dinucleotide (NAD) (final conc. 1.75 mg/ml) to 10 ml polypep base solution. Adjust to pH 8.0 with 5 M NaOH. Add 30 mg NBT immediately prior to use.

2.3. Visualisation

1. Phosphate-buffered saline (PBS).
2. 4% Phosphate-buffered formalin.
3. Hydromount (National Diagnostics).
4. Axioplan microscope (Zeiss).
5. Zeiss filter set #10 (excitation BP 450–490 nm, beam splitter FT510 nm, emission BP 515–565 nm).
6. Imaging and analysis software, e.g. Axiovision software (Zeiss).

3. Methods

Cytoplasmic LDH activity is used to determine the viability of osteocytes within cancellous bone sections from human femoral heads, though it is possible to use any type of bone from any type of donor or donor site. Cancellous bone was processed to 250 μm thick, undecalcified, unfixed sections and used directly for LDH staining and further analysis. However, Wong et al. have shown that the method can be used on overnight decalcified bone sections as well (1). Furthermore, Noble and Stevens demonstrated the use of the LDH assay on undecalcified cryosections (5). Their method used a 40% polypep-based LDH solution for section stabilisation during staining. Adapted from this method, we recommend using a 5% polypep base solution for LDH activity determination of osteocytes within 250 μm thick sections of cancellous bone.

3.1. Preparation of Fresh Cancellous Bone Sections for LDH Assay

1. Human femoral heads will be used as cancellous bone material for LDH viability assay. Therefore, cut the bone in 7-mm thick sections using an "Exakt 300" band saw. Prepare cancellous bone cores of 10 mm diameter using diamond tipped core bit. Cut the prepared cancellous bone cores to produce 250 μm sections using an annular saw. All blades need to be diamond-coated to enable cutting through the highly mineralised bone tissue. During all cutting and drilling procedures, the bone explants need to be irrigated with sterile

pre-cooled (4°C) saline to reduce the formation of bone debris and heat-induced cell death.

2. Collect the cancellous bone sections and place them individually in wells of 24-well plates. Sections need to be submerged in HBSS prior to LDH staining.

3.2. LDH Staining of Cancellous Bone Sections

1. As basis solution for the LDH assay, prepare a 5% polypep base solution (see Note 1). Polypep is a mixture of peptides, which dissolved in water results in a viscous solution stabilising tissue sections. To prepare 100 ml LDH solution, weigh out 0.66 g of the buffering substance glycylglycine and add 50 ml distilled or deionised water.
2. The solution is made isotonic by the addition of 0.75% sodium chloride (0.75 g/100 ml).
3. Adjust the pH of the solution with 1 M NaOH to pH 8.0.
4. Add 5 g polypep to the solution and leave it for 1–2 h to stir at room temperature, before adjusting to the final volume of 100 ml using distilled or deionised water. The prepared base solution might be stored up to 6 month if kept at 4°C.
5. The LDH assay solution is prepared fresh on the day of the assay from the pre-prepared base solution. Add 60 mM lactic acid as primary substrate to the solution. For a volume of 10 ml LDH solution, add 54 mg lactic acid. Mix the solution.
6. Add the second substrate, NAD^+, in a concentration of 17.5 mg per 10 ml to the solution and mix well. Both lactate and NAD^+ are substrates in the LDH catalysed reaction which results in NADH and pyruvate production due to enzyme activity.
7. Due to the addition of lactic acid, the pH of the LDH solution needs to be readjusted to pH 8.0 prior to the addition of the third and final substrate.
8. Add 30 mg NBT per 10 ml solution to the LDH solution. NBT is a light-sensitive substance, it is therefore necessary to protect the solution from light during the procedure. Mix the solution well to ensure complete dissolving of the NBT (see Note 2).
9. Cover cancellous bone sections with suitable volume of LDH staining solution. For a 250 μm thick section with 10 mm diameter, use a 24-well plate and 500 μl LDH solution for incubation. Allow the reaction to perform for 4 h at 37°C in the dark (see Note 3). During this process, active LDH present in viable cells converts the supplied substrates – lactate, NAD^+, and NBT – to produce pyruvate, NADH, and the visible purple formazan salt which is deposited inside the cytoplasm.

3.3. Documentation and Quantification of Viable Osteocytes Within Human Cancellous Bone

1. Wash off the LDH solution from the cancellous bone sections with several volumes of deionised water prior to a final wash with 50°C warm PBS.
2. Preserve LDH stained sections by the addition of 4% neutrally phosphate-buffered formalin at 4°C. Sections can be stored up to 6 months prior to analysis.
3. Prior to visualisation, wash the fixed sections shortly in deionised water. Place one section on a microscopic glass slide. Prior to the addition of Hydromount prepare a frame of nail polish in the size of the coverslip around the bone section. This will reduce the amount of air bubble formation after mounting. Cover the section with Hydromount. Apply a coverslip onto the section and leave slides for 2–4 h at room temperature.
4. Visualise the LDH stained sections using a fluorescence microscope. View slides at 20× lens magnification with Zeiss filter set #10 (excitation BP 450–490 nm, beam splitter FT510 nm, emission BP 515–565 nm) to visualise the dark-stained osteocytes blocking the green background autofluorescence of bone (see Note 4) (Fig. 2a). This provides an image that can be more easily analysed than brightfield images (Fig. 2b).
5. Take several micrographs in the centre of the sections. Analysis can then be performed on "zvi"-format micrographs using

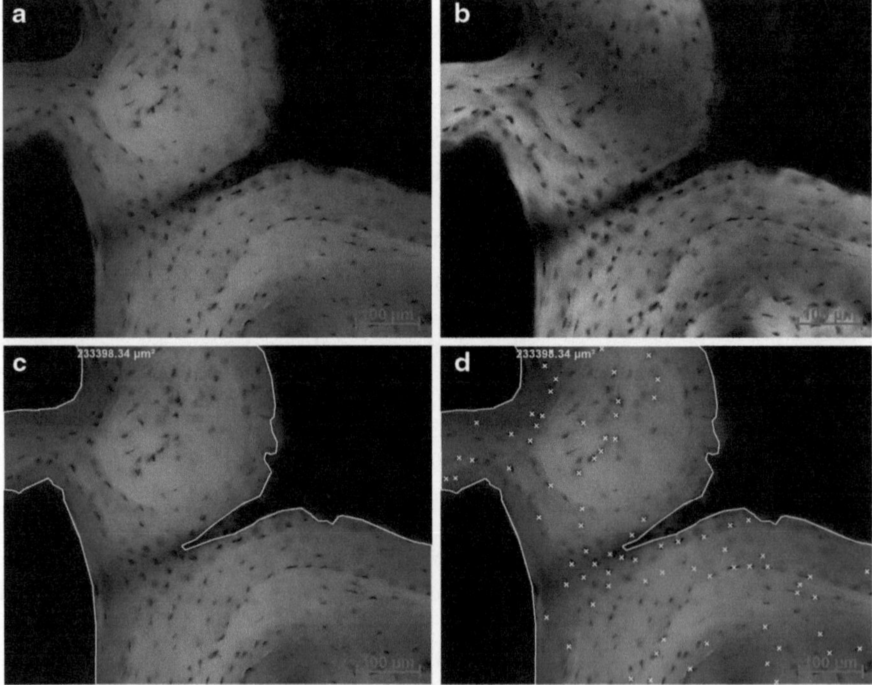

Fig. 2. Micrographs from LDH-stained human cancellous bone section. (**a**) Fluorescent micrograph (emission 515–565 nm) showing dark-stained osteocytes blocking the natural autofluorescent of the extracellular matrix of bone. (**b**) Brightfield micrograph of the same region of interest demonstrating an uneven bone matrix illumination in comparison to micrograph (**a**). Micrograph (**a**) was used for the determination of bone matrix area (**c**) and viable osteocytes (**d**).

the Axiovision software (Zeiss). Therefore, mark the extracellular bone matrix area per micrograph using the outline tool (Fig. 2c). Then mark the dark-stained osteocytes within this area (Fig. 2d). From this the ratio of viable osteocytes per mm^2 bone matrix area can be calculated (see Note 5).

4. Notes

1. The LDH viability assay on unfixed cryosections of bone tissue is possible. Therefore, prepare a 40% polypep base solution by adding 40 g of polypep instead of 5 g per 100 ml base solution. The addition of sodium chloride is not required for this solution.
2. The prepared LDH solution needs to be used immediately after the NBT is dissolved. If the colour of the prepared solution changes from yellow to a brown tone, we recommend to discard the solution and to prepare a fresh LDH solution.
3. Seal the sections during incubation with the LDH solution. The LDH solution is very viscous and may dry out quickly during incubation at 37°C.
4. For any LDH viability study we recommend to perform, next to a viable control, a "dead control" sample. Two methods have been used successful by our group to achieve consistent results – over night heat incubation at 56°C, as well as short cycles of repeated freeze (–196°C) and thawing (37°C). The LDH staining performed after 48 h of over night heat incubation demonstrates the presence of empty non-stained lacunae (Fig. 3).
5. To determine the ratio of live versus dead cells, stain the unprocessed bone tissue with ethidium homodimer prior to LDH staining. Dead cells will show a red-fluorescent nucleus.

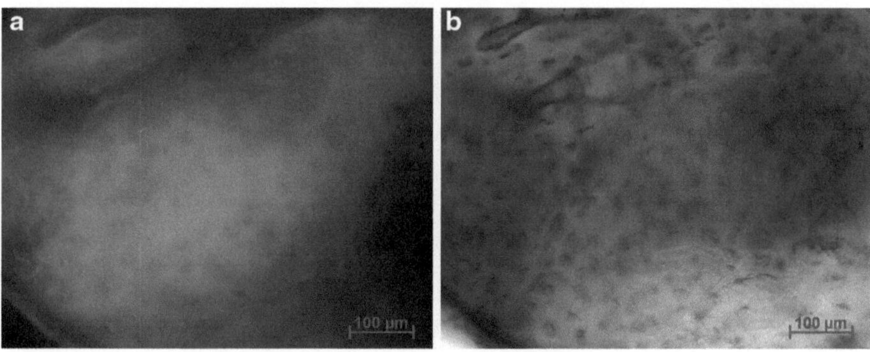

Fig. 3. Micrographs from sections of human cancellous bone heat-incubated over night at 56°C for 48 h prior to LDH staining. (**a**) Fluorescent micrograph (**a**; emission 515–565 nm) and brightfield micrograph (**b**) showing no positively stained osteocytes.

References

1. Wong, S.Y., Dunstan, C.R., Evans, R.A. & Hills, E. (1982) The determination of bone viability: a histochemical method for identification of lactate dehydrogenase activity in osteocytes in fresh calcified and decalcified sections of human bone. *Pathology* **14**, 439–442.
2. Dawson, D.M., Goodfriend, T.L. & Kaplan, N.O. (1964) Lactic dehydrogenases: functions of the two types rates of synthesis of the two major forms can be correlated with metabolic differentiation. *Science* **143**, 929–933.
3. Voet, D.J., Voet, J.G. & Pratt, C.W. (2002) Lehrbuch der Biochemie. Wiley-VCH.
4. Stoddart, M.J., Furlong, P.I., Simpson, A., Davies, C.M. & Richards, R.G. (2006) A comparison of non-radioactive methods for assessing viability in ex vivo cultured cancellous bone: technical note. *Eur. Cell Mater.* **12**, 16–25.
5. Noble, B.S. & Stevens, H.Y. (2003) Bone Research Protocols. Helfrich, M.H. & Ralston, S.H. (eds.), pp. 225–236 (Humana Press, Totowa).

Chapter 16

Measuring Glutamate Receptor Activation-Induced Apoptotic Cell Death in Ischemic Rat Retina Using the TUNEL Assay

Won-Kyu Ju and Keun-Young Kim

Abstract

Glutamate receptor activation-mediated excitotoxicity has been hypothesized to cause cell death in both acute and chronic neurodegenerative diseases including glaucoma. Although the precise mechanisms of ischemia-induced neuronal death are unknown, glutamate excitotoxicity-induced apoptotic cell death is considered to be an important component of postischemic damage in the retina. The blockade of apoptotic cell death induced by glutamate receptor activation provides strong evidence that glutamate excitotoxicity-induced apoptotic cell death may be a central mechanism of cell death in ischemic rat retina. We have shown that there is TUNEL-positive apoptotic cell death in the outer nuclear layer, inner nuclear layer, and ganglion cell layer of the ischemic rat retina at 12 h.

Key words: Retina, Ischemia, Glutamate receptor, Excitotoxicity, MK801, Apoptotic cell death, TUNEL

1. Introduction

Ischemia-induced damage of the retina is a relatively frequent event occurring in the course of a variety of pathological processes. This and the accessibility of the retina for manipulation of blood flow have promoted the use of different experimental models to investigate neuronal responses following ischemia–reperfusion injury (1–8). Since the retina is organized in discrete cell layers, this model also facilitates to study the effects of transient ischemia on different types of neurons. As in other central nervous system tissues, retinal ischemia results in apoptotic neuronal cell death. Most studies on the postischemic retina have reported that cell death is pronounced in the inner retina, which comprises the innermost ganglion cell layer (GCL) and the inner nuclear layer

Martin J. Stoddart (ed.), *Mammalian Cell Viability: Methods and Protocols*, Methods in Molecular Biology, vol. 740,
DOI 10.1007/978-1-61779-108-6_16, © Springer Science+Business Media, LLC 2011

(INL) containing the somata of amacrine, bipolar and horizontal cells, and those of the retina-specific Muller glia (6, 9–11). Although the precise mechanisms of ischemia-induced neuronal cell death are unknown, excitotoxicity triggered by the overactivation of glutamate receptors is considered to be a central component of postischemic damage in the retina (1, 12, 13). The preferential susceptibility of inner retinal neurons to ischemic damage would be in line with this concept, since expression of glutamate receptors is confined to neurons of the GCL and INL (14, 15).

Apoptosis is the most common form of eukaryotic cell death that can be classified by its morphological, biochemical, molecular, and functional aspects (16, 17). It has been reported that apoptosis is accompanied by rounding-up of the cell, retraction of pseudopods, reduction of cellular volume (pyknosis), chromatin condensation, nuclear fragmentation (karyorrhexis), classically little or no ultrastructural modifications of cytoplasmic organelles, plasma membrane blebbing, and engulfment by resident phagocytes, in vivo (18). Commercial availability of enzymatic in situ labeling of DNA strand breaks using terminal deoxynucleotidyl transferase (TdT), which catalyzes polymerization of labeled nucleotides to free 3′-OH DNA ends in a template-independent manner (TUNEL-reaction), and fluorescein labels provide a precise, fast, and simple nonradioactive technique to detect and quantify apoptotic cell death at single cell level in cells and tissues.

2. Materials

2.1. Transient Retinal Ischemia

1. 3-Month-old female Sprague–Dawley rat (200–250 g, Harlan Laboratories, Indianapolis, IN) (see Note 1).
2. Ketamine (100 mg/kg, Ketaset; Fort Dodge Animal Health, Fort Dodge, IA). Store at room temperature.
3. Xylazine (9 mg/kg, TranquilVed, St. Joseph, MO). Store at room temperature.
4. Sodium chloride is dissolved in sterilized water at 0.9%. Store at 4°C.
5. MK801 (Sigma). Working solution (10 mg/kg) is prepared by dilution in 0.9% saline. Store at 4°C.

2.2. Tissue Preparation

2.2.1. Fixation

1. Paraformaldehyde working solution (4%): 4 g Paraformaldehyde is dissolved in 60 mL of warmed (40–50°C), distilled water, add two drops of 0.1N NaOH, and adjust up to 100 mL with 5× PBS. Filter with #1 Whatmann paper. The working solution should be prepared freshly in each experiment.

2. 0.1N Sodium hydroxide. Store at room temperature.
3. Ice-chilled 5× and 1× phosphate-buffered saline (PBS). Store at 4°C.
4. 30-gauge needle and perfusion motor (L/S Analog Console Precision Pump Systems, Barnant Company, Barrington, IL).
5. Scissors, forceps, and 100 mm dish.
6. #1 Whatmann filter paper.

2.2.2. Wax Embedding

1. Ethanol is diluted with distilled water (50, 70, 80, 90, and 95%). Store at room temperature.
2. Cetyl alcohol. Store at 4°C (see Note 2).
3. 500 g Poly(ethylene glycol) (400) distearate (PEGD; Polysciences, Inc., Warrington, PA) is melted at 57°C and then mixed with 55.5 g of cetyl alcohol with stirring on hot plates at 60°C. Prepared pure wax can be labeled with pure wax I and II. Store at 4°C (see Note 3).
4. Intermediate wax solution is mixed with ratio 1:1 (100% ethanol: pure wax) (see Note 3).
5. Forceps.
6. 20 mL Disposable scintillation vials.
7. Cryomold Biopsy, disposable vinyl specimen molds (10 mm×10 mm×5 mm) (Tissue-Tek, Torrance, CA).

2.2.3. Gelatin Coating

1. 4.11 g Gelatin from porcine skin (type A) is dissolved in heated distilled water (800 mL, final 0.05%, 40–50°C) with stirring. Working solution should be freshly prepared for each slide coating. When the gelatin is thoroughly dissolved, chromium (III) potassium sulfate dodecahydrate (400 mg) is added with stirring. Once the gelatin working solution is thoroughly dissolved, the working solution is filtered with #1 Whatmann filter paper. Store at room temperature (see Note 4).
2. #1 Whatmann filter paper.
3. Microscope slides (Superfrost; Fisher Scientific) (see Note 5).

2.3. TUNEL Staining

1. Xylazine and ethanol.
2. 0.1% Sodium citrate solution is prepared by dilution with distilled water. Store at room temperature.
3. 0.15% Triton X-100 solution is prepared by dilution with 0.1% sodium citrate solution. Store at room temperature.
4. 20 mg/mL Proteinase K. The stock solution is prepared by dilution with 10 mM Tris/HCl, pH 7.4–8. Store at –20°C. The working solution (20 μg/mL) is prepared by dilution with 1× PBS.

5. In Situ Cell Death Detection Kit (Fluorescence; Roche Applied Science, Indianapolis, IN). Store at –15 to –25°C. Kit contents: enzyme solution (TdT from calf thymus, recombinant in *Escherichia coli*, in storage buffer, 10× conc.) and label solution (nucleotide mixture in reaction buffer, 1× conc.).
6. 3,000–3 U/mL DNase I recombinant in 50 mM Tris–HCl, pH 7.5, 1 mg/mL. Store at –20°C.
7. 1% BSA dissolved in PBS and aliquoted with 1 mL in 1.5 mL Eppendorf tube. Stable at –20°C for up to 1 year.
8. 10 mg/mL Hoechst 33342 (trihydrochloride, trihydrate) for nuclear staining. Working solution is freshly prepared by diluting 1:10,000 in PBS. Store at 4°C.
9. Fluoromount-G (Southern Biotech, Birmingham, AL) for antifade mounting. Store at room temperature.

3. Methods

3.1. Transient Retinal Ischemia

1. Female Sprague–Dawley rats, 3 months old (200–250 g in weight), are anesthetized with a mixture of ketamine (100 mg/kg) and xylazine (20 mg/kg) by intraperitoneal (IP) injection.
2. A cannula is inserted into the anterior chamber that was connected by flexible tubing to a reservoir as shown in Fig. 1.
3. By raising the reservoir, intraocular pressure is elevated above systolic blood pressure (100–120 mmHg) for 60 min. Animals are allowed to recover for 3–24 h.

3.2. Tissue Preparation

1. The light-adapted rats are anesthetized with isoflurane followed by an IP injection of a mixture of ketamine/xylazine as above.
2. Both eyes are enucleated and then the rats were killed by CO_2 inhalation.
3. The retinas are dissected from the choroid and fixed in 5 mL of 4% paraformaldehyde in 1× PBS (pH 7.4) for 4 h at 4°C.
4. After 1× PBS rinse, retinas are dehydrated through 5 mL of graded series of ethanol (50, 70, 80, 90, 95, and 100%) at room temperature, immersed in intermediate wax solution, pure wax I solution, and pure wax II solution at 37°C. The retinal tissues are then embedded in pure wax and placed at room temperature. The wax blocks for retinal tissues are stored at 4°C.
5. The wax blocks for retinal tissues are cut by 5–7-μm thickness using microtome (Reichert-Jung 2030; McBain Instruments, Chatsworth, CA).

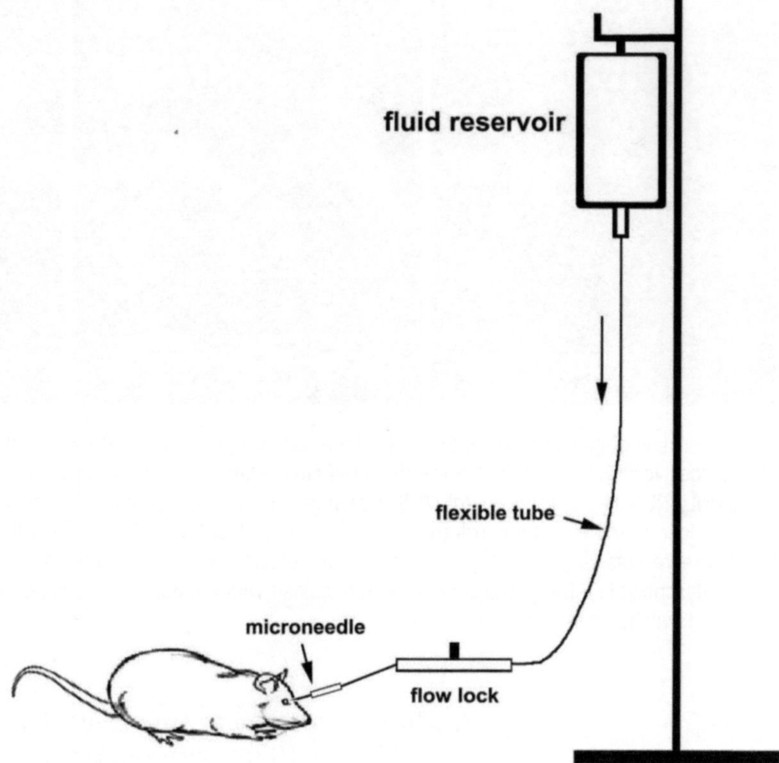

Fig. 1. Diagram of the instrument system used to elevate acute intraocular pressure.

6. The retinal sections are attached onto gelatin-coated glass slides using the water floating method (see Note 6).
7. The slides that have retinal tissues are stored at 4°C until use.

3.3. TUNEL Staining

1. The retinal tissue sections are dewaxed with xylene, and rehydrated with a graded series of ethanol (100, 95, 90, 80, and 70%) and distilled water. Each step is processed for 5 min at room temperature.
2. After 1× PBS rinse for 10 min, the sections are incubated with proteinase K working solution (20 μg/mL in PBS) for 7–10 min at 37°C (see Note 7).
3. The sections are rinsed with 1× PBS three times for 5 min.
4. Prepare a TUNEL reaction mixture with 10 μL of enzyme solution (vial 1) and 490 μL of label solution (vial 2) and mix well to equilibrate components at room temperature (see Note 8).
5. Each section is incubated with 50 μL of a TUNEL reaction mixture in a humidified chamber for 60 min at 37°C (see Note 9).
6. The sections are rinsed with 1× PBS three times for 5 min.

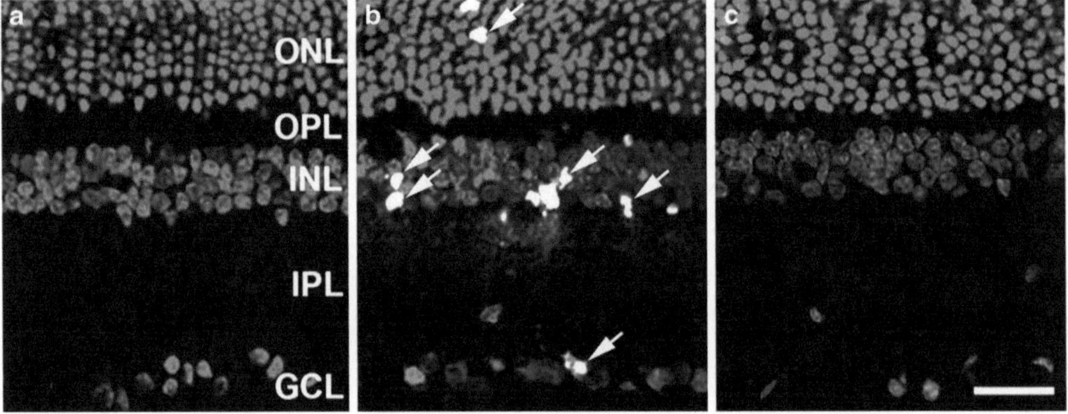

Fig. 2. Blockade of glutamate receptor activation blocks apoptotic cell death in ischemic rat retina. **(a)** There were no TUNEL-positive cells in the normal retina. **(b)** In ischemic retina of rats receiving vehicle, apoptotic cell death was present in the ONL, INL, and GCL (arrows). **(c)** MK801 treatment prior to ischemia blocked apoptotic cell death in all layers. ONL outer nuclear layer, OPL outer plexiform layer, INL inner nuclear layer, IPL inner plexiform layer, GCL ganglion cell layer. Size bar represents 20 μm **(a–c)**. The images were obtained on a Nikon Eclipse fluorescence microscope using a 40× lens as described (1). The sections were counterstained with the nucleic acid stain Hoechst 33342/PBS. (Reproduced from ref. 1 with permission from Molecular Vision).

7. The sections are incubated with 1 μg/mL Hoechst 33342 in 1× PBS for 10 min at room temperature to stain the nuclei.
8. The sections are rinsed with 1× PBS three times for 5 min.
9. The sections are mounted with Fluoromount-G and stored at 4°C until imaging.
10. The images are captured under fluorescence microscopy (Eclipse microscope, model E800; Nikon Instruments Inc., Melville, NY) equipped with a digital camera (SPOT; Diagnostic Instrument, Sterling Heights, MI). Excitation at 494 nm induces the Fluorescein (green emission) for the TUNEL-positive cells, while excitation at 350 nm induces Hoechst 33342 (blue emission). Software is used by Simple PCI version 6.0 software (Compix Inc., Cranberry Township, PA). An example of the signal for TUNEL-positive cells is shown in Fig. 2.

4. Notes

1. All procedures concerning animals were in accordance with ARVO Statement for the Use of Animals in Ophthalmic Vision Research and under protocols by institutional IACUC committees at the University of California San Diego. The Sprague–Dawley rats were housed in covered cages, fed with a standard rodent diet ad libitum, and kept on a 12-h light/12-h dark cycle.

2. Cetyl alcohol should be melted in an oven at 60°C overnight for mixing with PEGD.
3. PEGD (500 g) should be melted in an oven at 60°C overnight. Both pure wax I and II, and intermediate wax solutions can be melted at 50°C in the oven overnight for use in an experiment the following day. Both wax solutions can be reused. Stable at 4°C for up to 1 year.
4. The gelatin should not be added to distilled water that exceeds 50°C because high temperature can denature the gelatin.
5. Glass slides should be clean before coating. The slides can be placed in gelatin working solution for 2 min, removed from solution, covered by aluminum foil, and dried in refrigerator for 1 h. When slides are completely dried, they can be placed in gelatin coating solution for an additional two times, allowed to dry between each coating, and kept to dry completely in refrigerator for overnight. Stable at 4°C for up to 6 months.
6. Because the melting point of wax is 37°C, distilled water at room temperature should be used to spread and attach intact tissue sections onto glass slides.
7. The incubation time for proteinase K is important because proteinase K working solution can induce detachment of retinal tissues from coated glass slides. We have found 7 min is the best incubation time to preserve tissues on glass slides.
8. The TUNEL reaction mixture should be prepared immediately before use and kept on ice until use. The dilution rate for TUNEL reaction mixture will depend on the tissue samples. In retinal tissues with thickness of 7 μm, the recommended concentration from Kit was very strong and we have found that the best dilution rate of the TUNEL reaction mixture is 1 (enzyme solution) to 10 (label solution).
9. Two negative controls and a positive control should be included in each experimental setup according to In Situ Cell Death Detection Kit. For negative control, the sections are incubated with 50 μL of label solution instead of TUNEL reaction mixture. For positive control, the section is incubated with micrococcal nuclease or DNase I recombinant (3,000–3 U/mL in 50 mM Tris–HCl, pH 7.5, 1 mg/mL bovine serum albumin, Roche Applied Science). The samples should be put under aluminum foil and the room lights dimmed for subsequent steps.

Acknowledgment

This work was supported by National Institute Health grant R01 EY018658 (WKJ).

References

1. Ju, W. K., Lindsey, J. D., Angert, M., Patel, A., and Weinreb, R. N. (2008) Glutamate receptor activation triggers OPA1 release and induces apoptotic cell death in ischemic rat retina, *Mol Vis* **14**, 2629–2638.
2. Buchi, E. R. (1992) Cell death in the rat retina after a pressure-induced ischaemia-reperfusion insult: an electron microscopic study. I. Ganglion cell layer and inner nuclear layer, *Exp Eye Res* **55**, 605–613.
3. Nickells, R. W. (1996) Retinal ganglion cell death in glaucoma: the how, the why, and the maybe, *J Glaucoma* **5**, 345–356.
4. Rosenbaum, D. M., Rosenbaum, P. S., Gupta, H., Singh, M., Aggarwal, A., Hall, D. H., Roth, S., and Kessler, J. A. (1998) The role of the p53 protein in the selective vulnerability of the inner retina to transient ischemia, *Invest Ophthalmol Vis Sci* **39**, 2132–2139.
5. Joo, C. K., Choi, J. S., Ko, H. W., Park, K. Y., Sohn, S., Chun, M. H., Oh, Y. J., and Gwag, B. J. (1999) Necrosis and apoptosis after retinal ischemia: involvement of NMDA-mediated excitotoxicity and p53, *Invest Ophthalmol Vis Sci* **40**, 713–720.
6. Ju, W. K., Kim, K. Y., Hofmann, H. D., Kim, I. B., Lee, M. Y., Oh, S. J., and Chun, M. H. (2000) Selective neuronal survival and upregulation of PCNA in the rat inner retina following transient ischemia, *J Neuropathol Exp Neurol* **59**, 241–250.
7. Zhang, Y., Cho, C. H., Atchaneeyasakul, L. O., McFarland, T., Appukuttan, B., and Stout, J. T. (2005) Activation of the mitochondrial apoptotic pathway in a rat model of central retinal artery occlusion, *Invest Ophthalmol Vis Sci* **46**, 2133–2139.
8. Russo, R., Rotiroti, D., Tassorelli, C., Nucci, C., Bagetta, G., Bucci, M. G., Corasaniti, M. T., and Morrone, L. A. (2009) Identification of novel pharmacological targets to minimize excitotoxic retinal damage, *Int Rev Neurobiol* **85**, 407–423.
9. Szabo, M. E., Droy-Lefaix, M. T., Doly, M., Carre, C., and Braquet, P. (1991) Ischemia and reperfusion-induced histologic changes in the rat retina. Demonstration of a free radical-mediated mechanism, *Invest Ophthalmol Vis Sci* **32**, 1471–1478.
10. Hayreh, S. S., and Weingeist, T. A. (1980) Experimental occlusion of the central artery of the retina. IV: Retinal tolerance time to acute ischaemia, *Br J Ophthalmol* **64**, 818–825.
11. Osborne, N. N., Larsen, A., and Barnett, N. L. (1995) Influence of excitatory amino acids and ischemia on rat retinal choline acetyltransferase-containing cells, *Invest Ophthalmol Vis Sci* **36**, 1692–1700.
12. Lombardi, G., and Moroni, F. (1994) Glutamate receptor antagonists protect against ischemia-induced retinal damage, *Eur J Pharmacol* **271**, 489–495.
13. Choi, D. W. (1996) Ischemia-induced neuronal apoptosis, *Curr Opin Neurobiol* **6**, 667–672.
14. Brandstatter, J. H., Koulen, P., and Wassle, H. (1998) Diversity of glutamate receptors in the mammalian retina, *Vision Res* **38**, 1385–1397.
15. Brandstatter, J. H., and Hack, I. (2001) Localization of glutamate receptors at a complex synapse. The mammalian photoreceptor synapse, *Cell Tissue Res* **303**, 1–14.
16. Melino, G. (2001) The Sirens' song, *Nature* **412**, 23.
17. Wyllie, A. H., Kerr, J. F., and Currie, A. R. (1980) Cell death: the significance of apoptosis, *Int Rev Cytol* **68**, 251–306.
18. Kroemer, G., Galluzzi, L., Vandenabeele, P., Abrams, J., Alnemri, E. S., Baehrecke, E. H., Blagosklonny, M. V., El-Deiry, W. S., Golstein, P., Green, D. R., Hengartner, M., Knight, R. A., Kumar, S., Lipton, S. A., Malorni, W., Nunez, G., Peter, M. E., Tschopp, J., Yuan, J., Piacentini, M., Zhivotovsky, B., and Melino, G. (2009) Classification of cell death: recommendations of the Nomenclature Committee on Cell Death 2009, *Cell Death Differ* **16**, 3–11.

Chapter 17

Exploiting the Liberation of Zn^{2+} to Measure Cell Viability

Christian J. Stork and Yang V. Li

Abstract

Zn^{2+} ions are a critical component of cellular machinery. The ion is required for the function of many cell components crucial to survival, such as transcription factors, protein synthetic machinery, metabolic enzymes, hormone packaging, among other roles. In stark contrast to the cells' necessity for a sufficient Zn^{2+} supply, an excess of free Zn^{2+} is a situation that results in acute toxicity. Under normal conditions, free Zn^{2+} levels in the cell are extremely low; whereas estimates of free Zn^{2+} are in the subpicomolar range. In this way, the detection of elevated intracellular Zn^{2+} can be exploited as a highly sensitive and specific signal to indicate neuronal dysfunction. We have shown that the relationship between intracellular Zn^{2+} accumulation and the development of cellular injury/death to be ubiquitous among each of five tissue types tested; demonstrating the broad application and utility of the present technique.

Key words: Zinc, Cell death, Cytotoxicity, Fluorescence, Brain slice

1. Introduction

Zinc (Zn^{2+}) is required for the function of many enzymes, and plays a vital structural role in many other proteins (1). The levels of intracellular Zn^{2+} have been shown to be meticulously regulated, where free Zn^{2+} is essentially absent from the intracellular environment with estimates in the subpicomolar range (2, 3). This facet of Zn^{2+} homeostasis means that the detection of elevated intracellular Zn^{2+} serves as a very sensitive marker of neuronal dysfunction.

The neuronal accumulation of Zn^{2+} released into intracellular sources upon injurious stimuli is a well-documented phenomenon. Following transient forebrain ischemia, chelatable Zn^{2+} was shown to accumulate specifically in degenerating neurons in the hippocampus, cerebral cortex, and other brain regions (4). Apparent in more than just necrotic cell death, the increase of intracellular Zn^{2+} has also been reported in neurons undergoing

Martin J. Stoddart (ed.), *Mammalian Cell Viability: Methods and Protocols*, Methods in Molecular Biology, vol. 740, DOI 10.1007/978-1-61779-108-6_17,

apoptosis after target removal (5). The event of Zn^{2+} accumulation during cell death has even been reported in cultured cortical neurons exposed to chemically induced apoptosis (6). The evidence from multiple laboratories thus clearly documents the occurrence of Zn^{2+} elevation in the cellular signaling of apoptosis and the pathology of necrosis; demonstrating that the accumulation of intracellular Zn^{2+} occurs ubiquitously in the cascade of cellular demise (4–10). In this way, elevated intracellular Zn^{2+} specifically indicates the development of neuronal death. Furthermore, considering that the levels of intracellular free Zn^{2+} are so extremely low in healthy neurons (less than one ion of free Zn^{2+} per cell), the detection of elevated Zn^{2+} in neurons also serves as a highly sensitive means of detecting cell death.

While working with a rat hippocampus slice model of ischemia, researchers in our laboratory consistently observed a subset of neurons to be brightly labeled by the extracellular forms of fluorescent Zn^{2+} indicators. This was an unusual observation because the extracellular forms of these indicators are charged species and as such are normally excluded from the cell by the nonpolar plasma membrane (11, 12). It was hypothesized that the selectively labeled neurons were injured or dead and had lost membrane integrity; allowing the charged dye species to enter the newly permeable membrane to fluorescently label-free Zn^{2+}. Such locally elevated Zn^{2+} levels were seen to persist in dead and dying neurons for hours after injury. These results led us to conclude that Zn^{2+} was being liberated from intracellular sources throughout cell death, and that this event could be exploited as a means of labeling injured cells. To confirm that the cells being labeled by the Zn^{2+} indicator were indeed injured or dead, we employed the conventional cell death indicator propidium iodide (PI) (see Fig. 1). Overall, the results confirmed the detection of elevated intracellular Zn^{2+} to be both a specific and sensitive means of identifying neurons undergoing both apoptosis and necrosis (13).

The current method describes the preparation of brain slices for observation of neuronal injury by confocal fluorescence microscopy.

2. Materials

2.1. Animal Preparation

1. Male Sprague–Dawley rats weighing between 200 and 300 g are used as the source of tissue.
2. Ketamine HCl (10 ml/kg).
3. Cyanomethacrylate glue ("superglue").
4. Vibratome (Vibratome, Bannockburn, IL) equipped with a Vibratome 900R Media Cooling Unit (Vibratome, Bannockburn, IL) (bath temperature is maintained at 2–4°C).

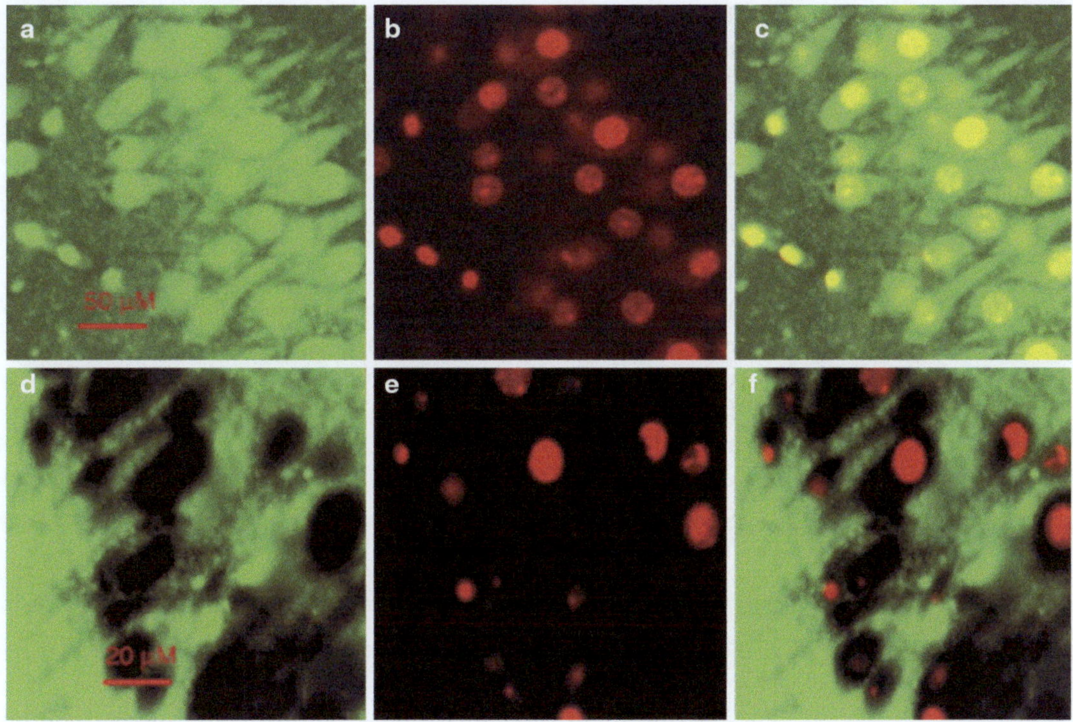

Fig. 1. Microscopic fluorescence images of rat hippocampal sections stained with NG and PI. (**a–c**) are images of pyramidal cells of a hippocampal CA1 section viewed at 100×. (**a**) NG stained cell bodies *green* with emission at 505 nm; (**b**) PI stained *red nuclei* with emission at 645 nm; (**c**) composite overlay fluorescence images showing colocalization of PI nuclei within cell bodies labeled with NG. (**d–f**) are images of pyramidal cells in the same type of preparation, but labeled with the live cell stain Calcein-AM in addition to PI. (**d**) The image of cells stained with Calcein-AM; (**e**) the PI stained nuclei of the same area; (**f**) the composite image showing anticolocalization of Calcein-AM labeled cells in apposition to PI-stained nuclei. (Reproduced from ref. 13 with permission from Elsevier Science).

2.2. Tissue Section (Slice) Preparation

1. Artificial cerebrospinal fluid (ACSF) prepared fresh: 121 mM NaCl, 1.75 mM KCl, 2.4 mM $CaCl_2$, 1.3 mM $MgCl_2$, 26 mM $NaHCO_3$, 1.25 mM KH_2PO_4, and 10 mM glucose; equilibrated with 95% O_2/5% CO_2.
2. Newport Green (NG, a zinc indicator. Invitrogen , Carlsbad, CA) is dissolved in dH_2O at 1 mM, stored in aliquots at −20°C, protected from light, and added to fresh ACSF at a concentration of 10 μM on the day of the experiment.
3. Propidium iodide is dissolved in dH_2O at 1 mg/ml (1.5 mM) and stored at 2–6°C, protected from light (PI is a potential mutagen and should be handled with, and disposed of, with care). Solutions are stable for 6 months. PI is diluted to a concentration of 5 μg/ml in fresh ACSF on the day of the experiment.
4. Permeabilizing solution for use as a positive control: 0.5% (v/v) Triton X-100 in ACSF.

2.3. Confocal Fluorescent Imaging System

1. Inverted Zeiss Axiovert LSM 510 Confocal Microscope (or other inverted microscope).
2. Zeiss 40× (1.3 NA) oil immersion, and 10× (0.3 NA) objective lenses (or other objectives).
3. Series RC-20 Small Volume Imaging Chamber (Harvard Apparatus, Holliston, MA) setup with appropriate perfusion lines (generally 1/8 ID tubing).
4. Peri-Star Pro 4-channel peristaltic pump (World Precision Instruments, Sarasota, FL) (or other peristaltic pump). Flow rate should be adjusted to 2 ml/min. Two of the channels are used to circulate the normal ACSF, where one channel drives fluid from the ACSF reservoir to the imaging chamber, and the other channel drives fluid from the chamber to a waste container, and the remaining two channels are set up to deliver the NG-ACSF and any experimental solutions.

3. Methods

The measurement of cellular injury by this technique is done by exploiting the progressive liberation of Zn^{2+} upon cell injury/death in addition to the concomitant loss of membrane integrity. After preparation, slices should sit for 30 min in ACSF (equilibrated with 95% O_2/5% CO_2) at room temperature to allow for recovery from slicing.

Tissue slices are examined while being perfused (2 ml/min) with ACSF (equilibrated with 95% O_2/5% CO_2) on the stage of an inverted confocal fluorescent microscope. Fluorescent recordings are made under baseline conditions (prior to dye addition) to determine any contribution by tissue autofluorescence.

3.1. Preparation of Tissue Samples

1. Male Sprague–Dawley rats weighing between 200 and 300 g are used as the source of tissue. Rats are anesthetized by intraperitoneal injection of ketamine HCl (1 mg/g body wt.) then decapitated and the brain is quickly removed and immersed in ice-cold ACSF (equilibrated with 95% O_2/5% CO_2) for 5 min. The cerebellum of the chilled brain is cut away in the coronal plane and then a single drop of cyanomethacrylate glue ("superglue") is used to mount the brain to the cutting platform. The mounted brain should have its anterior pole oriented upward, and the cutting blade should approach from the brain's superior portion to make coronal sections (orientation is important as initiating cuts from the inferior portion will destroy the brain tissue) (see Fig. 2). Once mounted, the brain is then placed into a prechilled ACSF bath of a Vibratome equipped with a Vibratome 900R Media Cooling Unit.

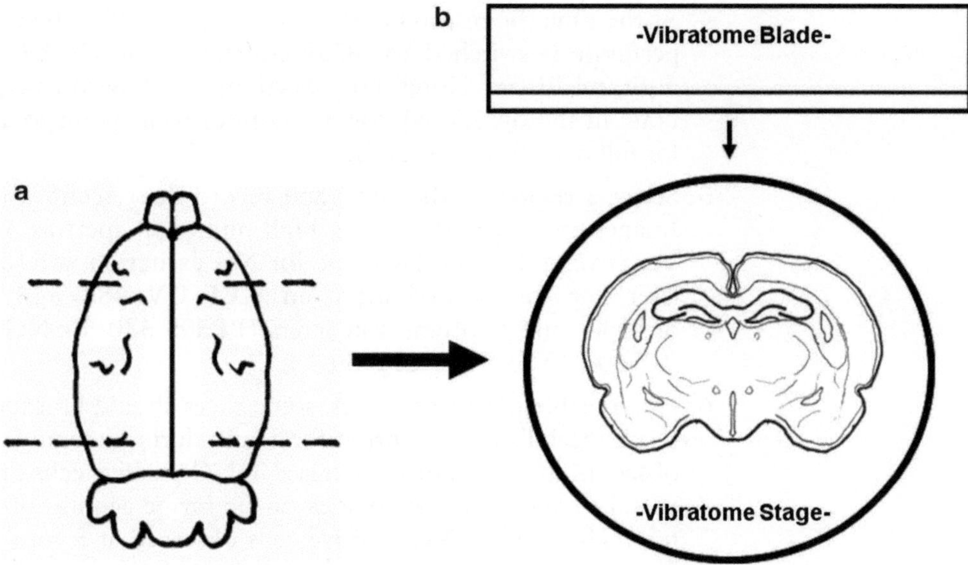

Fig. 2. The brain orientation for vibratome slicing is depicted. (**a**) Shows the top view of the rat brain. The *dashed lines* indicate where to cut the brain for mounting on the vibratome stage. After these cuts are made by hand, a drop of glue is placed on the vibratome stage and the brain is mounted on the vibratome stage as shown in (**b**) with the anterior facing upward. Coronal sectioning of the brain is then done, where the vibratome blade is made to approach the mounted brain as depicted in (**b**).

2. Tissue slices are made (sliced along the coronal plane) at 250 μm thickness (see Note 1), from which the hippocampi are isolated (see Note 2) and then transferred to an equilibration chamber with room temperature ACSF (bubbled with 95% O_2/5% CO_2). Slices are equilibrated for at least 30 min at room temperature prior to usage in order to promote recovery from the slicing procedure.

3.2. Confocal Fluorescence Imaging

1. On the stage of an inverted confocal fluorescent microscope, the tissue perfusion/recording chamber is secured and perfused with ACSF (bubbled with 95% O_2/5% CO_2) by a peristaltic pump at a constant rate (2 ml/min).
2. Following the equilibration period, hippocampus slices are placed into the recording chamber and secured with the tissue holder. Slices are perfused with ACSF constantly, and experimental agents are applied by dilution in ACSF where the perfusion intake line is switched from the normal ACSF to ACSF containing the experimental agent at the desired final concentration, and switched back to normal ACSF at the desired timepoint.
3. The tissue slices are treated with the desired experimental agents by switching the perfusate to ACSF containing the agent at its final concentration. After the desired exposure time, the perfusate is switched back to regular ACSF (see Note 3).

4. At the time the researcher desires to analyze cell viability, the perfusate is switched to ACSF containing 10 μM NG and 5 μg/ml PI (see Notes 4–6). Sections are allowed to equilibrate in the dye/ACSF for 5 min prior to imaging to allow for full dye saturation of the slice.
5. Select a region of the slice, and record fluorescent images. Images are recorded using a high numerical aperture (NA) objective, a 488-nm laser line for NG excitation, a 543-nm laser line for PI excitation, an HFT UV/488/543/633 dichroic, and two emission filters (BP505-530 for NG and BP560-615 for PI) (see Note 7).
6. As a reliable positive control, some slices should be exposed to permeabilization solution for 1 min during imaging. The observation of a dramatic increase in NG-positive cells can be useful to verify the capabilities of the image acquisition system. The level of NG-positive cells observed at 5 min after exposure of slices to the permeabilizing solution can be considered to represent a condition of maximum cell death.

4. Notes

1. When preparing tissue slices on a Vibratome, a good method for transferring newly cut slices from the Vibratome slicing bath to the equilibration chamber uses Q-tips; where the slice is picked up (in solution) by gently draping over the rounded cotton tip, and transferred to the new chamber.
2. In this work, the isolation and assay of neurons in hippocampus slices is highlighted. However, the present assay has been validated for use with tissue slices of the cerebral cortex, cerebellum, and kidney (see ref. 13).
3. During a switch between solutions, one must take care not to introduce air bubbles into the perfusion line intake, as air bubbles negatively impact image acquisition. One strategy to avoid bubbles during a solution switch is to briefly pause the peristaltic pump while the intake line is transferred between solutions. Another strategy is to pinch the intake line prior to its withdrawal from "solution 1" and release it once the line is fully submerged in "solution 2." Additionally, one other source of air bubble introduction can be a result of placing the perfusion intake line too close to the O_2/CO_2 outlet used for equilibrating the ACSF solution.
4. To minimize the amount of dye to be used, as little as 10 ml of dye solution will suffice if recirculated through the imaging chamber.

5. When employing this method with a cell type or tissue not previously documented, it is recommended that PI (5 μg/ml) be added to the NG solution and colocalization be demonstrated as a control. The addition of PI is otherwise not necessary for the current method to work.
6. Although the present protocol describes the usage of NG as the Zn^{2+} indicator, this method has been shown to work with at least two other extracellular Zn^{2+} fluorescent indicators; namely, ZinPyr-4 (Neurobiotex, Galveston, TX) and FluoZin-3 (Invitrogen, Carlsbad, CA) have also been demonstrated to function in a similar manner as NG (see ref. 13).
7. The settings for excitation laser intensities and detector settings should be determined with NG/PI-labeled tissue and then the same intensity settings should be tested with unlabeled tissue in order to ensure that there is negligible contribution of autofluorescence to the measured signal.

References

1. Vallee BL, Falchuk KH. The biochemical basis of zinc physiology. Physiol Rev 1993; 73(1):79–118.
2. Frederickson CJ, Bush AI. Synaptically released zinc: physiological functions and pathological effects. Biometals 2001; 14(3–4):353–66.
3. Colvin RA, Fontaine CP, Laskowski M, Thomas D. Zn2+ transporters and Zn2+ homeostasis in neurons. Eur J Pharmacol 2003;479(1–3):171–85.
4. Koh, J.Y., Suh, S.W., Gwag, B.J., He, Y.Y., Hsu, C.Y. & Choi, D.W. (1996) The role of zinc in selective neuronal death after transient global cerebral ischemia. Science, 272, 1013–1016.
5. Land PW, Aizenman E. Zinc accumulation after target loss: an early event in retrograde degeneration of thalamic neurons. Eur J Neurosci 2005;21(3):647–57.
6. Lee JY, Hwang JJ, Park MH, Koh JY. Cytosolic labile zinc: a marker for apoptosis in the developing rat brain. Eur J Neurosci. 2006 Jan;23(2):435–42.
7. Bossy-Wetzel E, Talantova MV, Lee WD, Scholzke MN, Harrop A, Mathews E, et al. Crosstalk between nitric oxide and zinc pathways to neuronal cell death involving mitochondrial dysfunction and p38-activated K+ channels. Neuron 2004;41(3):351–65.
8. Frederickson CJ, Koh JY, Bush AI. The neurobiology of zinc in health and disease. Nat Rev Neurosci 2005.
9. Hamatake M, Iguchi K, Hirano K, Ishida R. Zinc induces mixed types of cell death, necrosis, and apoptosis, in molt-4 cells. J Biochem (Tokyo) 2000;128(6):933–9.
10. Lobner D, Canzoniero LM, Manzerra P, Gottron F, Ying H, Knudson M, et al. Zinc-induced neuronal death in cortical neurons. Cell Mol Biol (Noisy-le-grand) 2000; 46(4):797–806.
11. Haugland HP. Handbook of fluorescent probes and research chemicals. Eugene, OR: Molecular Probes; 2001.
12. Ying H, Gottron F, Chen JW. Assessment of cell viability in primary neuronal cultures. In: Crawley JN, Gerfen CR, Sibley DR, Skolnick P, Wray S, editors. Current protocols in neurosciences. John Wiley & Sons Inc.; 2003.
13. Stork CJ, Li YV. Measuring cell viability with membrane impermeable zinc fluorescent indicator. J Neurosci Methods 2006;155(2):180–186.

Chapter 18

Noninvasive Bioluminescent Quantification of Viable Stem Cells in Engineered Constructs

Karim Oudina, Adeline Cambon-Binder, and Delphine Logeart-Avramoglou

Abstract

Bioluminescence from murine stem cells tagged with the luciferase gene reporter and distributed within three-dimensional scaffolds of two different materials is quantified in vitro and in vivo. The luminescence emitted from cells adhering to the scaffolds tested is monitored noninvasively using a bioluminescence imaging system. Monitoring the kinetics of luciferase expression via bioluminescence enables real-time assessment of cell survival and proliferation on scaffolds both in vitro and in vivo over prolonged (8 weeks) periods of time.

Key words: Bioluminescence imaging, Material scaffold, Noninvasive, Cell quantification, Cell survival, Cell proliferation, Tissue engineering, Real-time assessment

1. Introduction

Cell-based tissue engineering strategies rely on a combination of cells, material scaffolds, and techniques that enhance the survival and function of transplanted cells including tissue-progenitor cells. Current practice for the assessment of cell numbers within porous, three-dimensional (3D) matrices is a difficult task relying on methods that prove destructive to the sample for cell removal, collection, and analysis.

Cell assays that use reporter molecules whose activity can be measured, and often imaged, continuously in living cells have been developed. Combined with recent developments and availability of imaging systems of great sensitivity, these new technologies provide noninvasive methods to assess cellular and/or molecular processes within a live organism while preserving tissue

Martin J. Stoddart (ed.), *Mammalian Cell Viability: Methods and Protocols*, Methods in Molecular Biology, vol. 740,
DOI 10.1007/978-1-61779-108-6_18, © Springer Science+Business Media, LLC 2011

integrity in real time (1). Specifically, optical bioluminescence technology involving the genetic labeling of cells with a luciferase reporter-gene was successful in tracking viable cells on material scaffolds either prior to, or after, implantation in animals for tissue engineering purposes (2–6). Bioluminescence refers to the enzymatic generation of visible light through the oxidation of an enzyme (such as luciferase)-specific substrate in the presence of oxygen and, usually, adenosine triphosphate (ATP) as a source of energy. Photons emitted from cell-containing material constructs implanted in the animal model are detectable by ultrasensitive light detectors. A major advantage of using the luciferase enzymes as optical indicators in live mammalian cells and tissues is inherently low background, resulting from near absence of endogenous light emission from mammalian cells and tissues as well as from synthetic material scaffolds (1, 7).

The present assay quantifies the bioluminescence of murine stem cells tagged with the luciferase gene reporter and distributed inside scaffolds of either soft, translucent hydrogels or hard, opaque ceramic materials. Evaluation of luminescence emitted from cells adhering onto these material scaffolds is performed noninvasively using a bioluminescence imaging system. In vitro quantitation of photon fluxes (as an indicator of the number of viable luminescent cells) enables noninvasive quality control of viable cells seeded within constructs prior to implantation as well as real-time monitoring of cell survival, and proliferation in cell-containing constructs in vitro (for instance, in a bioreactor). In vivo quantitation of photon fluxes by measurements on the same animals over a period of time enables noninvasive and real-time monitoring of the fate of cells present on each synthetic material scaffold implanted in animals.

2. Materials

2.1. Preparation of Cell Constructs

1. Murine mesenchymal stem cells (C3H10T1/2) are genetically modified using lentiviral vectors encoding the firefly luciferase. A clonal line (referred to as C3H10-Luc) from these cells, exhibiting the highest luciferase expression, is selected and used for experiments (see Notes 1 and 2).
2. "Complete DMEM," consisting of Dulbecco's Modified Eagle's Medium (DMEM) supplemented with 10% fetal calf serum (FCS), penicillin (100 U/ml), and streptomycin sulfate (100 μg/ml).
3. Trypsin EDTA (1×) solution.
4. Disk-shaped AN69 hydrogels (diameter 6 mm; height 2 mm), with pores (mean diameter of 500–1,000 μm) and porosity of

77% are fabricated and kindly provided by Dr. Honiger, Saint-Antoine Hospital, Paris, France. The AN69 polymer is an acrylonitrile and sodium methallyl sulfonate co-polymer (8).

5. Cube-shaped ($3 \times 3 \times 3$ mm^3), porous [with mean diameter of 100–300 μm and porosity of 49% (Biocoral®, Inoteb, Inc., Levallois-Perret, France)] scaffolds of natural *Porites* coralline ceramic. The scaffolds consisted of 99% calcium carbonate in the form of aragonite (9, 10).
6. Peroxyacetic acid solution (Dialox™, Bioxal – S&M, Paris, France). A 10% Dialox™ solution in physiological saline (0.15 M NaCl) is used.

2.2. Implantation of the Cell Constructs

1. Nude mice, 9 weeks old, each one weighting 20–22 g (Janvier; Orleans, France) (see Note 3).
2. Ketamine (Ketalar®; Panpharma, Virbach, France).
3. Xylazine (Rompun® 2%; Bayer, France).
4. Povidone–iodine (Betadine®; Vetoquinol, France).
5. Ethicon nonabsorbable vicryl 3-0 sutures (Johnson and Johnson, Belgium).

2.3. Bioluminescence Acquisition

1. Phosphate buffer saline (PBS); pH 7.4.
2. DMEM without phenol red (w/o PR) supplemented with 10% FCS, penicillin (100 U/ml), and streptomycin sulfate (100 μg/ml), and 25 mM HEPES.
3. D-Luciferin, potassium salt (Interchim, Montluçon, France), stored at −20°C in the dark, dissolved in PBS (concentration of 20 mg/ml), and sterilized using 0.2-μm pore diameter filters. Either use this solution immediately or freeze aliquots at −20°C in the dark till future use (see Note 4). For in vitro imaging, prepare a solution (1,500 μg/ml in DMEM medium without phenol red; referred to as "5× luciferin solution"). For in vivo imaging, use a 20 mg/ml solution.
4. Isoflurane Belamont (Nicholas Piramal India Limited, London, England) liquid for inhalation by experimental animals.
5. IVIS Lumina bioluminescent imaging system (Xenogen – Caliper Lifescience, Trembley-en-France, France). The present system is equipped with a back-thinned, back-illuminated charge coupled device (CCD) camera cooled at −90°C (Peltier effect) with a (f/0.95–f/16) 50-mm lens and controlled by imaging software (Living Image® Software, version 3.1) (see Note 5). Imaging may be performed using other bioluminescent imaging systems.

3. Methods

In order to assess, either in vitro or in vivo, the viability of cells on constructs, transgenic cells expressing *luc*, the *firefly* luciferase gene (see Note 6), under the control of a constitutively active promoter (such as CMV or MLV promoters) must be prepared.

Using viral transduction (e.g., rMLV-LTR-Luciferase retroviral vectors), most of the cell types used in tissue engineering, including mesenchymal stem cells, can be stably transduced with the luc gene reporter. Following genetic labeling, the luminescent stem cells are distributed inside the scaffold materials to obtain cell-containing constructs. The two biomaterial scaffolds used in this study are (1) a soft, translucent polymeric hydrogel and (2) a hard and opaque ceramic scaffold. Upon addition of the substrate luciferin, the light generated by the photoprotein expressed inside the luminescent cells, and then transmitted through the scaffolds and the animal tissues can be monitored externally using the CCD camera. Scaffolds without cells provide the negative controls for light emission from the cell-containing constructs. Emitted photons are integrated by the camera over time; the signal intensity is reported as "photon flux." Despite signal attenuation (due to the presence of the scaffold material) observed in comparison to the signal obtained with cells in suspension (not on the scaffold), the luminescence signal still correlates with the number of cells present on each material scaffold tested, both in vitro and in vivo (see Note 7) (6).

3.1. Preparation of Cell Constructs

1. The AN69 hydrogels are sterilized via immersion in 10% (v/v) Dialox™ in physiological saline for 24 h followed by thorough rinsing using sterile, physiological saline. Each scaffold is then placed in individual 15-ml propylene tubes.
2. Coralline ceramic scaffolds are sterilized by autoclaving (at 121°C for 30 min). Prior to cell seeding, these coral scaffolds are immersed and kept in DMEM (containing 10% FCS) under slight vacuum for 30 min. Then, each scaffold is placed in individual 15-ml propylene tubes.
3. The C3H10-Luc cells are first rinsed with PBS, then treated with trypsin–EDTA (5 ml/175 cm^2 cell culture flask) at 37°C for 5 min. Complete DMEM (15 ml/175 cm^2 cell culture flask) is added, the cells are counted, collected by centrifugation (300 × *g* for 5 min), and are resuspended in DMEM w/o PR (see Note 8) (but containing 10% FCS, antibiotics, and 25 mM HEPES) to obtain a range of cells (specifically, 1×10^6, 2×10^6, and 3×10^6 cells/ml) in suspension. Aliquots (50 µl) of these cell suspensions (containing either 5×10^4, 10^5, or 1.5×10^5 C3H10-Luc cells) are delivered onto each material scaffold. The constructs thus obtained are maintained first

under a slight vacuum at room temperature for 15 min, and then under standard cell-culture conditions for 4 h. At that time, 2 ml of DMEM (w/o PR/10% FCS/Antibiotics/HEPES) are added and the cell constructs are maintained under standard cell culture conditions for 24 h before either in vitro bioluminescence analysis (see Subheading 3.3.2) or in vivo implantation (see Subheading 3.2).

3.2. Implantation of the Cell Constructs

1. The European Guidelines for Care and Use of Laboratory Animals (EEC Directives 86/609/CEE of 24.11.1986) are observed during all aspects of the animal studies.
2. Anesthetize mice by intraperitoneal injection of Ketalar (1 mg/10 g of body weight) and xylazine (0.1 mg/10 g of body weight). Disinfect the surgical implantation site (mouse back) using a povidone–iodine solution.
3. Using sterile surgical instruments, perform small incisions on both sides of the vertebral axis of each animal, create pouches between the muscles and the aponeurotic layer, and carefully insert the cell constructs (two per mouse; see Notes 9 and 10). The soft tissues at the implantation sites are closed with interrupted nonresorbable sutures.
4. Image the animal (at day 0) (see Subheading 3.3.2) immediately after surgery. At that time, the animals are still anesthetized (see Note 11).
5. Place each animal back in its cage near a heater (at low-to-medium setting) and carefully observe the animals till full and successful recovery from the surgical procedure and related anesthesia. Inject subcutaneously sterile physiological saline as needed to avoid dehydration of the animals.

3.3. Bioluminescence Acquisition

The instructions that follow are pertinent to the use of an IVIS Lumina bioluminescent imaging system. These comments can be adapted, however, using other bioluminescence imaging systems.

3.3.1. Parameters Setting

1. Initialize the imaging system and check the temperature of the camera (which should be in the −90°C to −105°C range depending on the imaging system used in the study).
2. Select the field of view; set the size of the stage area to be imaged by adjusting the position of the stage and lens.
3. Take a photograph of the subject(s) inside the chamber before photon acquisition and adjust the position of the subject(s) on the stage as needed.
4. For the first acquisition of a luminescent subject, select the "Auto-exposure," which automatically sets the exposure time, binning, and lens aperture to keep the signal within an optimal range. When not available, adjust the following settings manually:

(a) *Exposure time* is the length of time that the shutter of the camera is opened during acquisition of the luminescent image. Since the observed signal level is directly proportional to the exposure time, adjustment of the exposure time to obtain a signal above the noise [that is, above 600 counts with the IVIS Lumina equipment (see Note 12)], but less than the CCD camera saturation (approximately 60,000 counts). Integration times less than 0.5 s are not recommended (due to the finite time required to open and close the lens shutter); there is little benefit in using exposure times greater than 10 min.

(b) *Binning* (low, medium, or high) controls the pixel size on the CCD camera. Increasing the binning increases the pixel size and the signal-to-noise ratio (sensitivity), but reduces the spatial resolution.

(c) The *lens aperture* controls the amount of light detected and the depth of field. Usually, luminescent images are taken with a large aperture (f/1) to maximize sensitivity.

5. The biodistribution kinetics of the luciferin substrate within the cell-containing implanted constructs depends on the type (e.g., hydrogel versus ceramic, extent porosity, etc.) of the scaffold material as well as on the mode of delivery (local, intravenous, or intraperitoneal) of the luciferin substrate. For these reasons, each new in vitro and in vivo experiment requires a preceding kinetic study which is ***absolutely*** necessary in order to determine the lapse-time after which the peak of light emission appears. To this goal, acquire a sequence of images from one cell-containing construct with constant setting parameters starting immediately after the luciferin substrate addition to the subject. Keep the determined lapse-time (between luciferin injection and photon acquisition) constant for all measurements taken during the study.
6. Once determined, exposure conditions (specifically, aperture, stage position, binning, and time after luciferin injection) are kept identical for all measurements during an experiment.

3.3.2. In Vitro Imaging

1. Under sterile conditions, place each cell-containing scaffold (constructs) into individual wells of a 96-well, black plate containing 200 μl of DMEM w/o PR/10% FCS/Antibiotics/HEPES (see Note 13).
2. Add 50 μl of D-luciferin 5× solution per well. Adjust the final volume of the luciferin-containing medium in each well in such a way that each construct is completely immersed. The final concentration of luciferin is 300 μg/ml (see Notes 4 and 14).
3. Immediately image the constructs-containing plate using the medium-binning setting of the instrument for an exposure

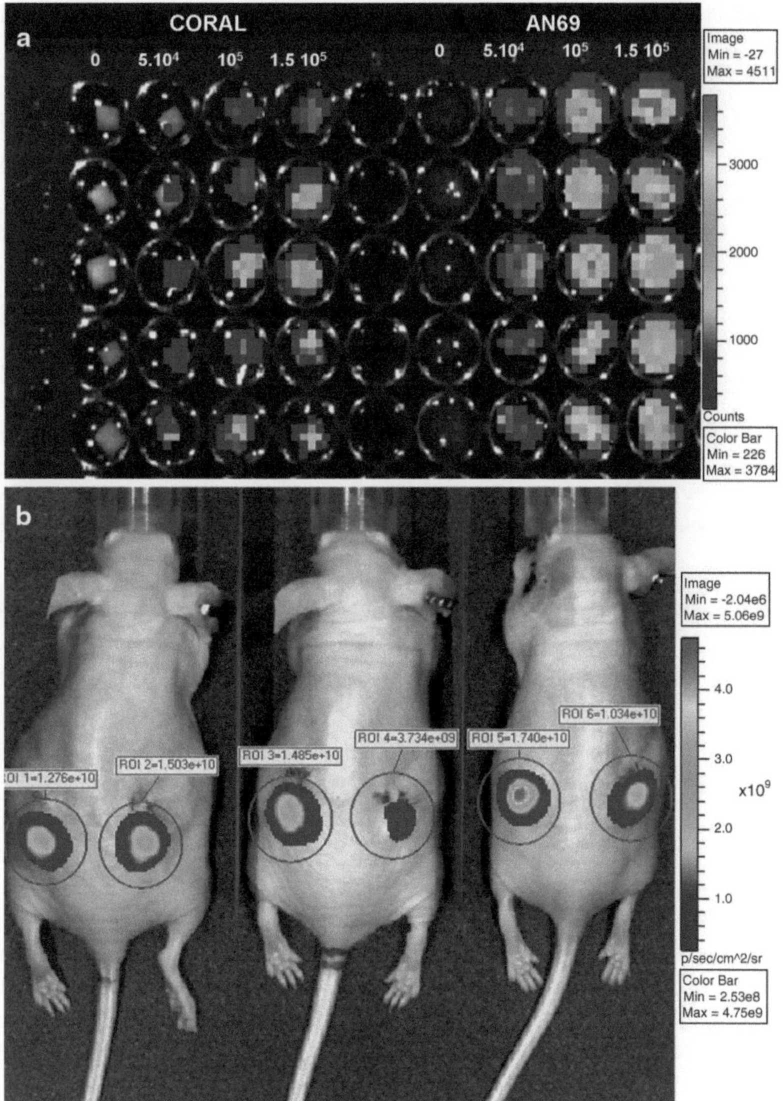

Fig. 1. Representative bioluminescent images using the IVIS Lumina BLI system. (**a**) In vitro bioluminescent image of a 96-well black plate containing either coral scaffolds or AN69 hydrogels either unseeded or seeded with 5×10^4, 1×10^5, or 1.5×10^5 C3H10-Luc cells. (**b**) In vivo bioluminescent image of C3H10-Luc cell-containing constructs subcutaneously implanted in the back of Nude mice (3 days postimplantation). *Circles* represent the standard regions of interest (ROI) manually delineated surrounding each implant. Data from each ROI are reported as "photon fluxes" that represent the emitted photon counts per second per centimeter square per steradian (p/s/cm²/sr).

time of 5 min (see Notes 15 and 16). An image acquisition of cell construct-containing plate is shown in Fig. 1a.

4. To remove the substrate, aspirate the luciferin solution and rinse the scaffolds with fresh complete medium three times under sterile conditions. Cell constructs are ready for either further in vitro tissue culture or in vivo implantation.

3.3.3. In Vivo Imaging

1. Place three mice in an induction chamber and anesthetize them using 4% isoflurane in 50% oxygen and 50% air, each at a flow rate of 2 l/min.
2. Inject an aliquot (50 μl) of D-luciferin (20 mg/ml) subcutaneously in each mouse in the area surrounding each implant (see Notes 4, 14, and 17).
3. Place the three mice (they are monitored at the same time) in the chamber of the imaging system in the prone position, and place their noses in a nose cone so that the mice continue to inhale isoflurane (1% with 1 l/min of O_2/air) and thus remain anesthetized (see Notes 18 and 19). Ten minutes after the D-luciferin injection, image the animals using the medium-binning setting for a 10-min exposure (see Notes 15 and 20). The acquisition time may be reduced to 1 min when the obtained images are over saturated (see Note 21). An image acquisition of cell construct-containing mice is shown in Fig. 1b.
4. Remove the animals from the chamber of the imaging system and place them back in their cages; carefully observe all animals till full and successful recovery from the anesthesia.
5. In order to monitor in vivo the fate (specifically, survival, proliferation, and death) of cells on the implanted constructs for long (8 weeks) periods of time in the same animal, image the animals at day 0 (day of implantation) and then twice-a-week for an 8-week period postimplantation (see Note 11).

3.3.4. Bioluminescence Analyses

Light emission is quantified using Living Image software. Each frame depicts the bioluminescent signal in pseudocolor, which is generated using an arbitrary color palette representing standard light intensity levels with blue being the least and red the most intense level, super-imposed on the respective gray-scale photographs.

1. Manually delineate standard regions of interest (ROI) surrounding each implant on the bioluminescent images. Data from each ROI are reported as "photon fluxes" that represent the emitted photon counts per second per centimeter square per steradian (p/s/cm^2/sr).
2. Subtract the signal background (obtained from measurements using scaffolds without cells) from the respective measurements from cell-containing constructs. Examples of pertinent in vitro and in vivo results are shown in Fig. 2 and examples of the time-courses of the in vivo cell proliferation on each material scaffold implanted in animals are shown in Fig. 3.

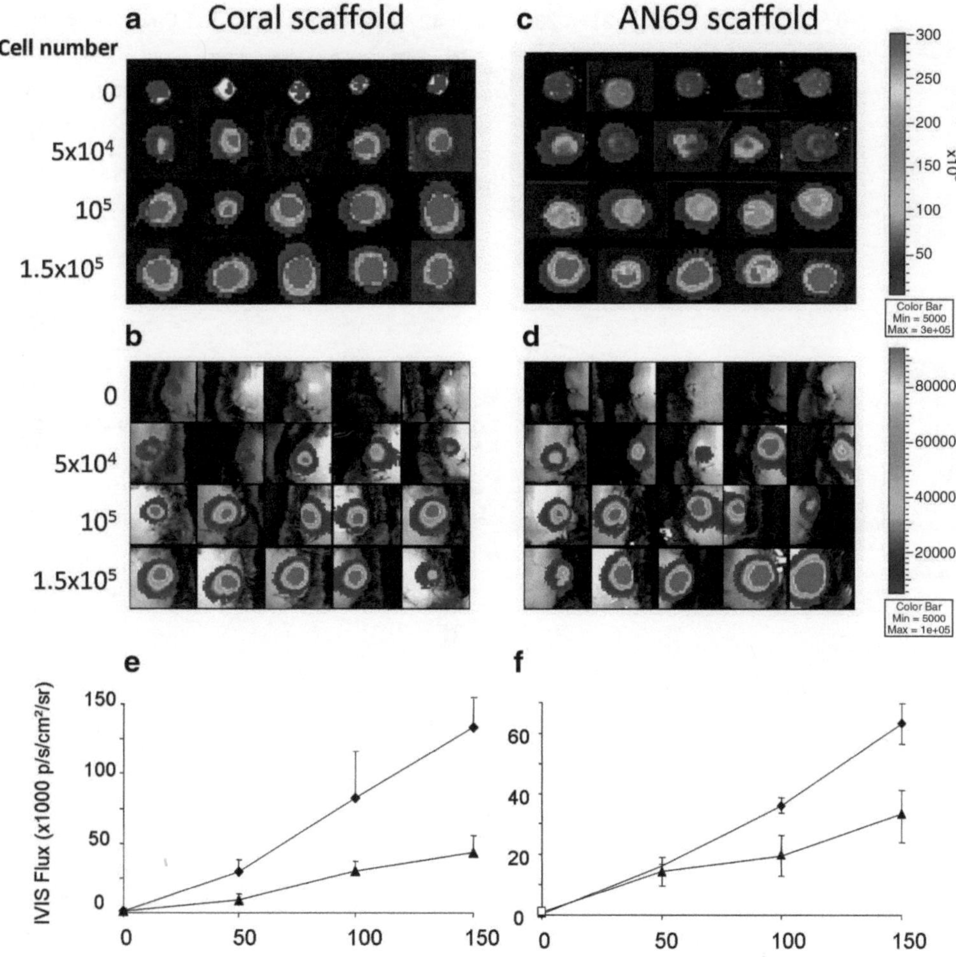

Fig. 2. In vitro and in vivo bioluminescence detection in C3H10-Luc cells on scaffolds using the IVIS Lumina BLI system. Representative bioluminescent images of 5×10^4 to 1.5×10^5 C3H10-Luc cells adhering to either coral scaffolds (**a**, **b**) or AN69 hydrogels (**c**, **d**) (each $n=5$) are shown in this figure. The cells were imaged before implantation in mice (**a**, **c**) and in vivo the day of implantation (**b**, **d**). The rainbow pseudocolor scale (ranging from 5×10^3 to 3×10^5 and from 5×10^3 to 10^5 p/s/cm²/sr for in vitro and in vivo, respectively) was adjusted for all images. The results show the bioluminescent signal (p/s/cm²/sr) from C3H10-Luc cells on either coral scaffolds (**e**) or AN69 hydrogels (**f**) in vitro before implantation (*black diamonds*), in vivo the day of implantation (*black triangles*). Values are average ± SD; $n=5$ for each condition tested (from (6), © 2009 Mary Ann Liebert, Inc., publishers. Reprint with permission).

4. Notes

1. The results obtained from either the in vitro or in vivo studies quantified the bioluminescence signal from many (six logarithmic dilutions of luminescent) cells in order: (1) to determine the threshold of detected luminescent cells (a minimum of 30 C3H10-Luc cells could be detected above background in vitro)

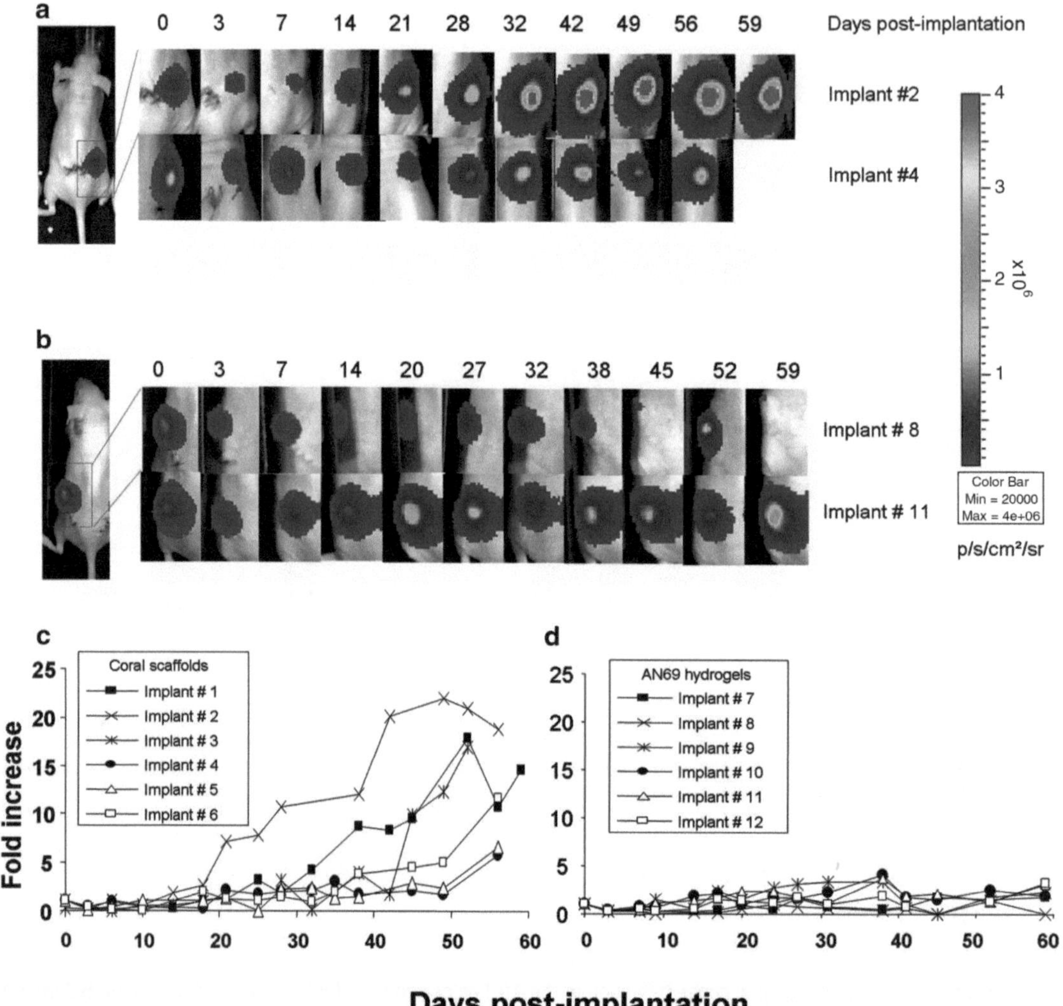

Fig. 3. Time-course of the in vivo proliferation of C3H10-Luc cells on scaffolds during 59 days of implantation. These are representative BLI images of cell proliferation in subcutaneously implanted coral scaffolds (**a**) and AN69 hydrogels (**b**). Two implants were chosen for each scaffold material tested and illustrate the lowest (specifically, implants #4 and #8 for the coral scaffolds and the AN69 hydrogels, respectively) and highest (implants #2 and #11 for the coral scaffolds and the AN69 hydrogels, respectively) cell proliferation results obtained in the study. The rainbow pseudocolor scale (ranging from 5×10^3 to 3×10^5 p/s/cm²/sr) was adjusted for all images. After subtracting the signal background of empty scaffolds, the photon fluxes (p/s/cm²/sr) quantified from BLI images of C3H10-Luc cells on either coral scaffolds or AN69 hydrogels were normalized to the respective values obtained on the day of implantation of each seeded scaffold and were plotted versus days postimplantation (**c, d**) (from (6), © 2009 Mary Ann Liebert, Inc., publishers. Reprint with permission).

and (2) to establish the correlation between the emitted signal and the cell number (this correlation was linear in the range of 10–10^6 C3H10-Luc cells).

2. In vitro assess the stability of *luc* expression in genetically modified cells as a function of cell passage. In addition, assess the stability of the *Luc* expression in multipotent progenitor cells during their differentiation toward lineage committed cells.

3. Light is transmitted more efficiently through the tissues of white or hairless (such as nude) mice because melanin (of the skin pigment) absorbs substantial amounts of light and because light is also scattered significantly by dark animal fur. When using animals with dark fur, it is recommended to use a depilatory cream and shave the fur from the body region in which the cell-containing constructs are implanted.
4. Because D-luciferin is both light and oxygen sensitive, it should be stored as small aliquots in vials covered with foil. Repeated freeze–thaw is not recommended.
5. Imaging may be performed using other bioluminescent imaging systems.
6. The *firefly* luciferase enzyme is widely used for in vivo BLI because it emits light over a broad spectrum which peaks at 560 nm and includes a significant fraction of light above 600 nm. The emitted bioluminescent light is not absorbed by hemoglobin within the aforementioned region of the spectrum and is, therefore, much more readily transmitted through biological tissues (1). Another luciferase used in studies with mammalian cells is the *Renilla* luciferase, which emits blue light with a peak of 480 nm; the substrate of this enzyme is coelenterazine. Therefore, *firefly* and *Renilla* luciferases may be distinguished spectrally. However, because the *Renilla* luciferase emits blue light, and biological tissues exhibit high adsorption within this region of the spectrum, this enzyme has had limited in vivo applications (11).
7. Transmission of light through synthetic material scaffolds (as well as through mammalian tissues) is affected by both light scattering and absorption (1). Scattering occurs when light hits cell and cytoplasmic organelle membranes. Light absorption depends both on the type of the synthetic material scaffolds (within which luminescent cells are seeded) (6) and on the surrounding tissues (in which hemoglobin, melanin, and other pigment macromolecules may be present).
8. The presence of phenol red decreases the transmitted light signal because the red color absorbs part of the emitted photons.
9. To limit light absorption by surrounding tissues in vivo, a superficial implantation site is preferable.
10. Ensure separation of the two implanted cell-containing constructs to minimize interference of the transmitted bioluminescence signals from multiple samples (a minimum of 10 mm is recommended).
11. Variations in the number of cells seeded onto each scaffold are accounted for when reporting cell survival as a function of time by normalizing the photon fluxes obtained over time to

those obtained at the beginning of the study. The accuracy of the first measurement is, therefore, critical. We have compared the BLI signals from cell-containing constructs acquired immediately after implantation (the number of implanted cells is identical to the number of seeded cells but the substrate may leak from the incision site while the postsurgery trauma milieu may affect light emission) to those acquired at day 1 (postsurgery trauma is negligible but the cell numbers may have already changed). The best accuracy (evaluated from correlations between the in vitro signals before implantation and the in vivo signals postimplantation) is obtained when the animals were imaged on day 1 postsurgery.

12. These counts are proportional to the number of photons detected per pixel.
13. Because some plastic materials (such as translucent polystyrene) strongly reflect and scatter incident light, opaque (either black or white) plates should be used instead. In addition, opaque plates avoid interference of signals emitted from different construct-containing wells. An alternative approach is to position black partitions between the construct-containing wells of a translucent plate.
14. Since the concentration of luciferin (used in excess to optimize the enzymatic reaction) slightly affects the level of the bioluminescent signal, the same luciferin concentration should be used for all experiments.
15. A number of parameters, such as the extent of labeling by the *luc* gene reporter, the cell number, etc., affect the level of the bioluminescent signal and, therefore, the exposure time during the acquisition of the luminescent image.
16. There is no time lapse between addition of the D-luciferin substrate and sample imaging because results of time-course experiments established that the maximum bioluminescence signal occurred within the first 2 min after the luciferin addition.
17. Because vascularization of the implanted constructs does not occur before 2 weeks postimplantation, the luciferin substrate is administrated locally (close to the implants) and is not injected systemically.
18. Place black partitions between two mice positioned laterally next to one another.
19. Care should be taken to position each mouse with the constructs containing the luminescent cells in such a way that the camera can detect the maximum emitted photons. For instance, the BLI signal from implants positioned laterally in the mouse will be underestimated because the light source will radiate with an angle to the vertical axis and to the camera, which is located at the top of the chamber.

20. The light emission kinetics were previously determined using cell-containing scaffolds implanted in mice, locally injected with luciferin, and sequentially imaged for an acquisition time of 5 min, at 5-min intervals for a total of 20 min (6). The optimal bioluminescence signal occurred at the interval of 10–15 min postadministration of luciferin in the animals.
21. Light emission from an implant displaying a very strong BLI signal may overlap with emission with the weak signal from an implant on the other side of the same animal; in this case, the weak signal will be overestimated. In order to measure accurately the light emission from the implant with the weak signal, administer the luciferin substrate locally and acquire the signal from the two implants sequentially starting with the weaker signal. It should be noted, however, that such a protocol is not applicable if the luciferin substrate is administered systemically; in that case, it is still possible to shield the implant with the highest signal using black cardboard or a similar material.

Acknowledgments

The authors gratefully acknowledge Dr. Jiri Honiger (Service de chirurgie orthopédique, Saint-Antoine Hospital, Paris, France) for donating the AN69 polymer scaffolds and Inoteb Inc. (Levallois-Perret, France) for donating the coral scaffolds used in this study. The authors also thank the Plate-Forme d'Imagerie Dynamique at the Pasteur Institute, Paris, France, for their assistance with bioluminescent imaging and Professor Rena Bizios and Dr. Ines Sherifi for critically reading the manuscript. This project was supported by the Centre National de la Recherche Scientifique and by a grant from the "Gueules cassées" Foundation.

References

1. Contag CH, Bachmann MH. (2002)Advances in in vivo bioluminescence imaging of gene expression. *Annu Rev Biomed Eng* 4, 235.
2. Blum JS, Temenoff JS, Park H, Jansen JA, Mikos AG, Barry MA. (2004)Development and characterization of enhanced green fluorescent protein and luciferase expressing cell line for non-destructive evaluation of tissue engineering constructs. *Biomaterials* 25, 5809.
3. Roman I, Vilalta M, Rodriguez J, Matthies AM, Srouji S, Livne E, Hubbell JA, Rubio N, Blanco J. (2007)Analysis of progenitor cell-scaffold combinations by in vivo non-invasive photonic imaging. *Biomaterials* 28, 2718.
4. Degano IR, Vilalta M, Bago JR, Matthies AM, Hubbell JA, Dimitriou H, Bianco P, Rubio N, Blanco J. (2008)Bioluminescence imaging of calvarial bone repair using bone marrow and adipose tissue-derived mesenchymal stem cells. *Biomaterials* 29, 427.
5. Giannoni P, Scaglione S, Daga A, Ilengo C, Cilli M, Quarto R. (2009)Short-time survival

and engraftment of bone marrow stromal cells in an ectopic model of bone regeneration. *Tissue Eng Part A.*

6. Logeart-Avramoglou D, Oudina K, Bourguignon M, Delpierre L, Nicola MA, Bensidhoum M, Arnaud E, Petite H. (2009) In vitro and in vivo bioluminescent quantification of viable stem cells in engineered constructs. *Tissue Eng Part C Methods.*
7. de Wet JR, Wood KV, DeLuca M, Helinski DR, Subramani S. (1987)Firefly luciferase gene: structure and expression in mammalian cells. *Mol Cell Biol* 7, 725.
8. Honiger J, Sarkis R, Baudrimont M, Delelo R, Chafai N, Benoist S, Sarkis K, Balladur P, Capeau J, Nordlinger B. (2000)Semiautomatic macroencapsulation of large numbers of porcine hepatocytes by coextrusion with a solution of AN69 polymer. *Biomaterials* 21, 1269.
9. Guillemin G, Meunier A, Dallant P, Christel P, Pouliquen JC, Sedel L. (1989)Comparison of coral resorption and bone apposition with two natural corals of different porosities. *J Biomed Mater Res* 23, 765.
10. Guillemin G, Patat JL, Fournie J, Chetail M. (1987) The use of coral as a bone graft substitute. *J Biomed Mater Res* 21, 557.
11. Bhaumik S, Gambhir SS. (2002) Optical imaging of Renilla luciferase reporter gene expression in living mice. *Proc Natl Acad Sci U S A* 99, 377.

Chapter 19

Raman Micro-Spectroscopy as a Non-invasive Cell Viability Test

Sophie Verrier, Alina Zoladek, and Ioan Notingher

Abstract

The number of techniques to identify, quantify and characterise cell death is rapidly increasing as more is known about the complex mechanisms underlying this process. However, most of these techniques are invasive and require preparation steps such as cell fixation, staining or protein extractions. Non-invasive analysis of living cells represents a key point in cell biology, e.g. in toxicology studies or in tissue engineering. In this chapter, we report the usefulness of Raman spectroscopy as a non-invasive method to distinguish cells at different stages of cell cycle and living cells from dead cells. Throughout two examples, we show the performance and the use of Raman spectroscopy as a new non-invasive method to assess cell viability.

Key words: Raman spectroscopy, Cell death, Apoptosis, Non-invasive

1. Introduction

Apoptosis is a natural cellular process known as programmed cell death. This process involves many biochemical modifications within the cells, leading to morphological changes, such as shrinkage of the cell, loss of cell junctions, membrane blebbing, chromatin condensation and relocalisation at the nuclear periphery (1).

Three major components of the cell are affected during apoptosis: DNA, proteins and lipids. The DNA double strand undergoes cleavage at the linker regions between nucleosomes, leading to multiple fragments of 150–200 bp (1). Apoptosis can induce protein degradation, modifications or activation (e.g. poly(ADP-ribose) polymerase, caspases) (2, 3). Lipids are also heavily involved in apoptosis, as the asymmetric distribution of the membrane phospholipid phosphatidylserine (facing the cytoplasmic compartment in healthy cells) is disturbed in apoptotic cells. This leads to the

Martin J. Stoddart (ed.), *Mammalian Cell Viability: Methods and Protocols*, Methods in Molecular Biology, vol. 740,
DOI 10.1007/978-1-61779-108-6_19, © Springer Science+Business Media, LLC 2011

exposure of the phosphatidylserine, the signal for phagocytic removal of apoptotic cells (4).

On the other hand, necrotic cell death is described in a more passive way. It is mostly characterised by the absence of apoptotic parameters.

The number of techniques to identify and quantify cell death, particularly via apoptosis is rapidly increasing as more is known about the complex mechanisms, which underlie this process. However, most of these techniques are invasive and require preparation steps such as cell fixation, staining or protein extractions.

Raman-spectroscopy is a well-established analytical tool and is based on the interaction of electromagnetic radiation with the sample molecules. A Raman spectrum represents a chemical fingerprint of the sample. A classic Raman set-up is shown in Fig. 1. Typically, a near-infrared laser is directed via a series of mirrors through a microscope onto the sample, resulting in light scattering. As a result of the interaction with the sample, a small portion of the scattered photons have a shifted frequency. After filtration of the non-shifted photons, only the photons with altered frequencies are directed to the detector. Each chemical bound results in a defined frequency shift of the photons. For example a Raman shift of 1,005 cm^{-1} correspond to the phenylalanine, the intensity of the peak is relative to the concentration (5). This technique has three main advantages over conventional biological assays when applied to the study of living cells (6): (1) it is rapid (1–3 min), (2) no labels are required so it is non-invasive, and (3) no damage is induced to the cells if suitable laser wavelength and intensities are used.

Raman spectroscopy has proven to be able to detect in situ and real-time biological changes related to cell cycle and cell death (7, 8). By measuring the magnitude of the 782–788 cm^{-1} Raman band of DNA, the distribution of DNA has been imaged in apoptotic cells, showing the fragmentation of the nucleus (9). The application

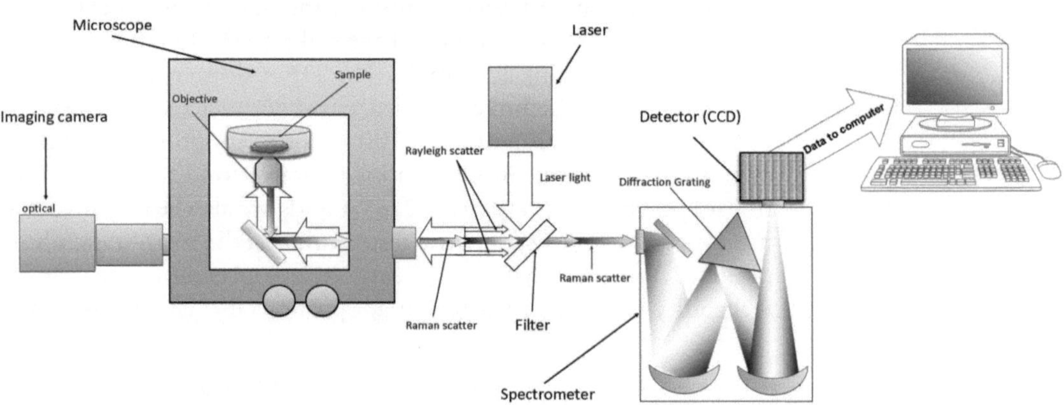

Fig. 1. Simplified schematics of a Raman System.

of Raman micro-spectroscopy as a detection system in a cell-based integrated biosensor has also been investigated (8).

In this chapter, we explored the use of Raman micro-spectroscopy as a non-invasive method to monitor cell death and provide two examples. We induced cell apoptosis using Triton X-100 and etoposide treatment and analysed cell death using Raman micro-spectroscopy. Even though not reported in the present chapter, and in order to establish a correlation between spectral changes and molecular changes within the cell, established biological methods (Western blotting, DNA integrity DNA-ladder kit and immunocytochemistry) were used in parallel.

2. Materials

2.1. Cell Culture

1. Human epithelial lung carcinoma cells (A549 cell line) American Type Culture Collection (ATCC, USA).
2. Human breast cancer cells (MDA-MB-231 cell line) European Collection of Animal Cell Culture (ECACC).
3. Ham's F-12 Kaighn's Modification cell culture medium (F12-K) supplemented with 10% foetal calf serum (FCS).
4. MEM culture medium supplemented with 10% DMSO, 30% foetal calf serum and 1% L-glutamine at 37°C and 5% CO_2.
5. Penicillin–Streptomycin 100 U/mL.
6. Sterile solution of trypsin (0.25%) and ethylenediamine tetraacetic acid (EDTA) 1 mM.
7. Phosphate Buffer Saline at pH 7.4 (PBS).
8. Polystyrene plastic culture dishes for cell amplification (see Note 1).
9. MgF_2 cover slides (Global Optics, UK).
10. Fused silica cover-slips.
11. 24-well plates.

2.2. Cell Treatment

Cell death was induced with the following products diluted as described in the corresponding example.

1. Triton X-100 (example 1).
2. Etoposide (VP16, Sigma) (example 2).

2.3. Raman Spectroscopy: Instrumentation

1. Raman spectrometer's major components (Commercial systems: Renishaw inVia, Horiba LabRam Inv) (see Note 2).
   (a) Laser as excitation source (see Note 3)
   (b) Wavelength selectors (see Note 4)

   (c) Samples illumination system and light collections optics (see Note 5)
   (d) Detector. Usually photosensitive CCD arrays are used to detect the Raman spectrum
   (e) A computer with appropriate software for collection of data
2. Principal Component Analysis (PCA) software (Matlab Statistical toolbox, Mathworks).
3. GRAMS/32 software (Thermo Galactic).
4. For long-lasting experiments, full incubator enclosure or heated stage can be assembled to the whole system (Solent Scientific, UK).

3. Methods

The Typical Raman spectrum of a live cell is shown in Fig. 2. The Raman spectrum consists of peaks corresponding to vibrational frequencies of various cellular components. A more detailed peak assignment can be found in reference (6).

However, Raman spectroscopy applied to cell biology questions is still at early stage of development. Depending on the application, a pre-processing is required for the baseline correction and normalisation of the intensity, the measurement's set-up, and the data analysis methods have to be adapted (see Note 6). In this paragraph, we describe the methods through two examples.

3.1. Example 1: Assessment of Apoptosis for Adherent Cells

1. Upon arrival, human epithelial lung carcinoma cells (A549) are grown as monolayer in F12-K medium supplemented with 10% FCS (37°C, 5% CO_2 humidified atmosphere).
2. When cells reach 85% of confluence, A549 are detached from the surface using trypsin–EDTA and seeded at the density of 15,000 cells/cm^2 on fused silicate substrate that are placed in wells of a 24-well plate. F12-K 10% FCS medium is added and cells are cultured for 3 days as mentioned in step 1.
3. For cell death induction, the cell culture medium is removed and replaced by treatment medium consisting in F12-K 10% FCS containing 100 μM Triton X-100.
4. Control cells receive normal cell culture medium and are analysed in the same way.
5. Raman spectroscopic measurements of the cells treated with 100 μM Triton X-100 are performed after 0, 24, 48, and 72 h of treatment (Fig. 3). WST-1 cell viability assay and DNA integrity assay are performed in parallel to control cell apoptosis.

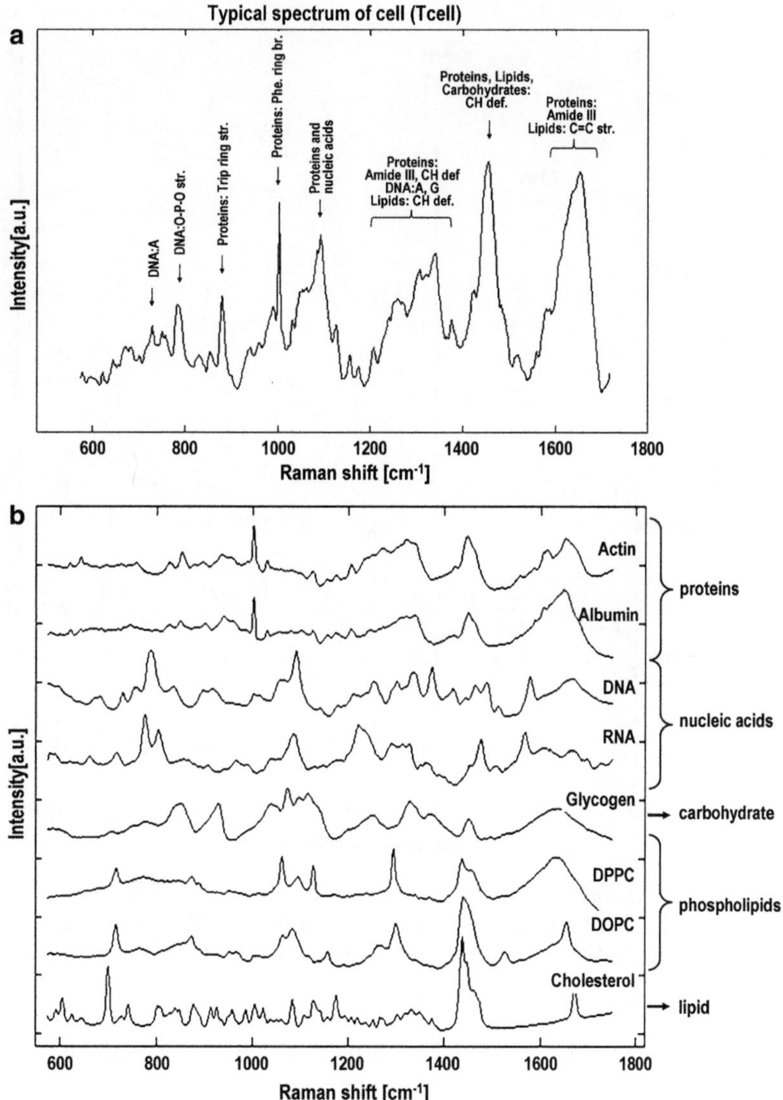

Fig. 2. (**a**) Typical average spectrum of a cell with major peak assignment. (**b**) Spectra of major biomolecules found in cells.

6. Raman spectra of cells are measured by 785-nm diode laser using a Renishaw RM2000 with Leica microscope and immersion objective (Leica 63 × NA = 0.9, 2 mm working distance).
7. Since the treated cells are round and reduced in size, the spectra are measured by focusing the laser spot (6 μm × 2 μm line) on the centre of the cell.
8. Ten cells, randomly chosen are measured at each time point.
9. An average Raman signal from the entire cell is obtained by the use of expanded (in planar plate) laser focal spot. In this method the average acquisition times per cell are usually in the range of tens of seconds.

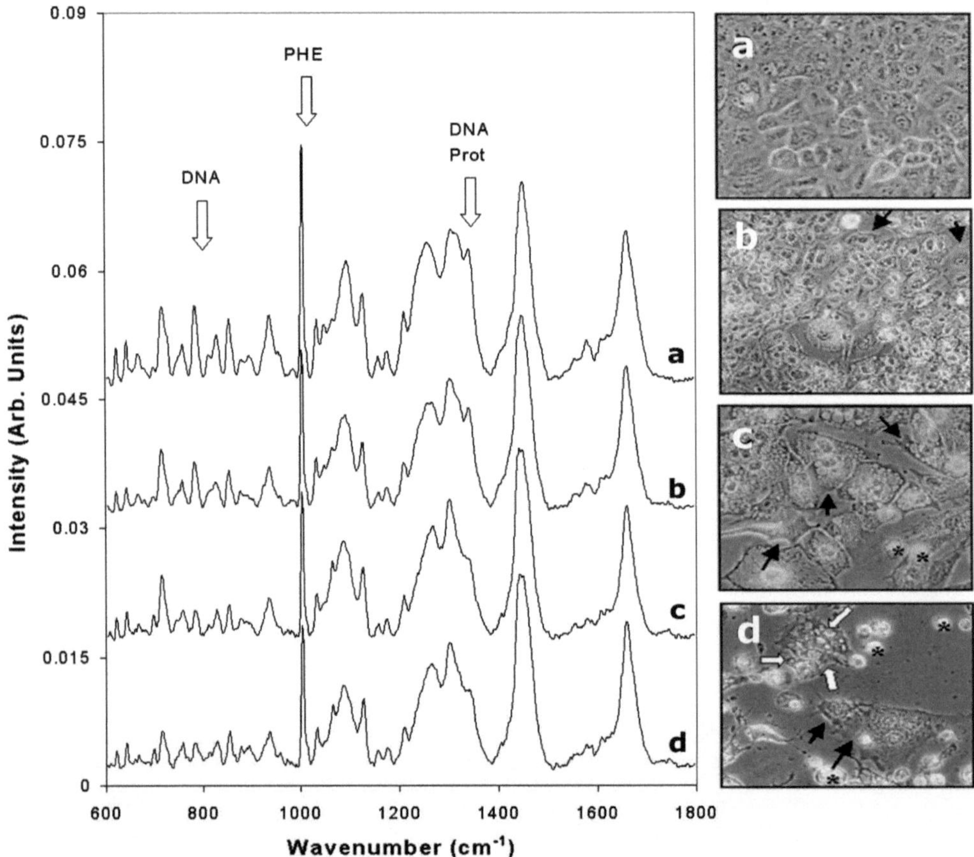

Fig. 3. Raman spectra of A549 cells treated with 100 μM Triton X-100 with corresponding light transmission microscopy pictures. (**a**) 0 h, (**b**) 24 h, (**c**) 48 h, and (**d**) 72 h post treatment (*asterisks*: floating cells, *black arrows*: membrane retraction, *white arrows*: cellular membrane convolution) (adapted from Verrier et al. (10) with permission from John Wiley and Sons).

10. The signal of the fused silica and cell culture media are measured using the same set-up.
11. All Raman spectra are processed: contributions of fused silica substrate and media are subtracted, each spectrum is normalised using the area of the 1,449 cm^{-1} band (C–H vibrations corresponding to all cellular components), and a multipoint baseline is used.
12. The area of individual vibrational bands was computed using the curve-fitting option of GRAMS/32 software (Thermo Galactic).
13. Raman spectra of the treated cells indicate significant biochemical changes related to DNA, proteins, and lipids (Fig. 4). Changes associated with the DNA structure and integrity is indicated by an 80% decrease in the 788 cm^{-1} peak (phosphodiester bonds O–P–O) after 48 h, which was confirmed by DNA integrity assays. A decrease in the magnitude of Raman

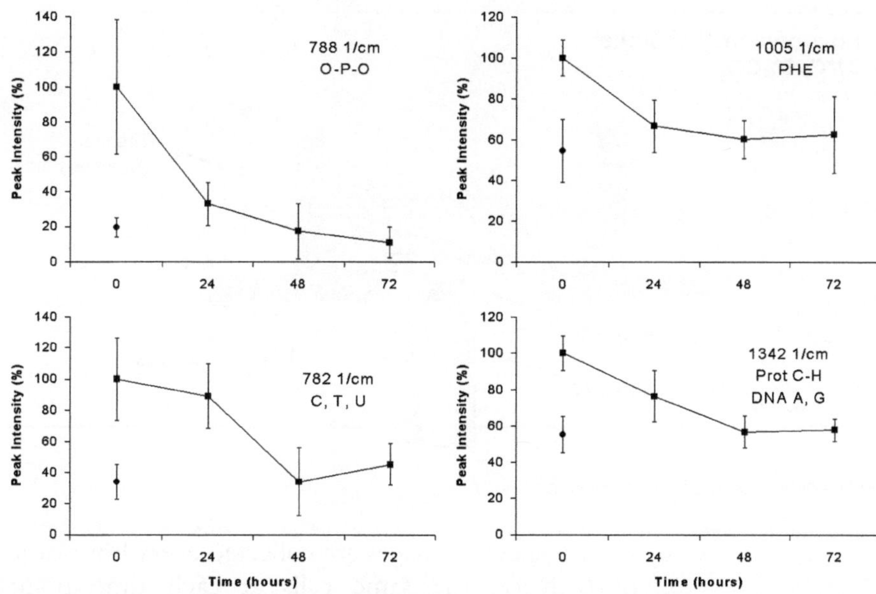

Fig. 4. Variation of DNA and protein Raman peaks of A549 cells as a function of 100 μM Triton X-100 exposure time (filled square: measured values, filled circle: values corresponding to dead cells) (adapted from Verrier et al. (10) with permission from John Wiley and Sons).

peaks corresponding to proteins, such as 1,342 cm^{-1} (44%) and 1,005 cm^{-1} (45%), is explained by protein cleavage following caspase activation. An increase in lipid concentration is found, indicated by the increase in lipid contribution to the 1,660 cm^{-1} band and by the appearance of the ester peak at 1,743 cm^{-1} band. The increase in lipid concentrations suggested production of apoptotic bodies (10).

3.2. Example 2: Time-Course Study of Apoptosis

1. Upon arrival, human breast cancer cells (MDA-MB-231 cell line) are grown in MEM culture medium supplemented with 10% DMSO, 30% foetal calf serum and 1% L-glutamine at 37°C and 5% CO_2.
2. For a time course experiment requiring analysis of the same cells, Raman microscope is equipped with an environmental enclosure (Solent Scientific, UK).
3. For Raman micro-spectroscopy measurement, 10,000 cells/cm^2 are seeded on a 0.17-mm thick MgF_2 coverslip, fitted in a customised made sterile titanium chambers with open bottom window (coverslip holder) as shown in the Fig. 5 and grown until reaching 50% confluence.
4. For cell treatment, the growth medium is replaced by etoposide-containing medium (300 μm in full cell growth medium).
5. During the whole experiment, cells are maintained in a controlled atmosphere at 37°C and 5% CO_2 suitable for cell culture (see Fig. 5).

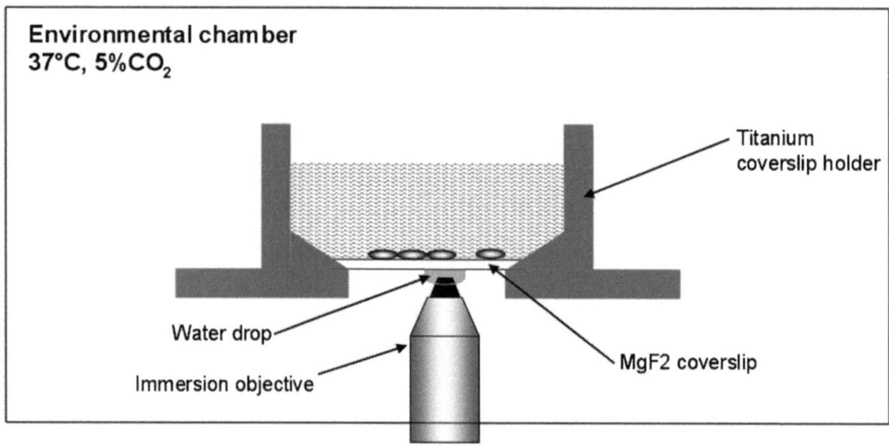

Fig. 5. Coverslip holder, schematic representation.

6. Raman spectral images are collected every hour over a period of 6 h on the same cells at each time point of the experiment.
7. The measurements are carried out on a home-build inverted Raman microscope (IX71, Olympus, UK) with a high precision step-motor stage (Prior, Cambridge, UK) using a 785-nm laser line for excitation of the Raman spectra (XTRA, Toptica Photonics, Munich, Germany).
8. Spectrometer is equipped with 830 lines/mm grating (spectral resolution of ~1.5 cm^{-1} in the 600–1,800 cm^{-1} region) and water-cooled deep-depletion CCD (Andor Technologies, Belfast, UK).
9. A water-immersion 60× objective, (NA 0.90, Olympus) is used for focussing the laser beam on individual cells as well as for collection of the Raman scattered photons.
10. Raman spectra are collected from ~1 µm regions of cell by using a diffraction-limited laser spot. The cell is scanned in a raster pattern with small step increments.
11. Step size is 1 µm and acquisition time for each pixel 1 s.
12. All analysis processes were performed in Matlab (Mathworks).
13. Background spectra are removed from the data set by the given threshold. Pre-processing of the spectra included smoothing, baseline correction and normalisation of spectra.
14. Images were obtained by calculating the ratio between the peak intensity corresponding to lipids (C=C symmetric stretching at 1,670 cm^{-1}) and proteins (phenylalanine breathing mode 1,003 cm^{-1}).
15. Images are constructed from point spectra collected at different spatial positions within a cell. Such high spatial resolution

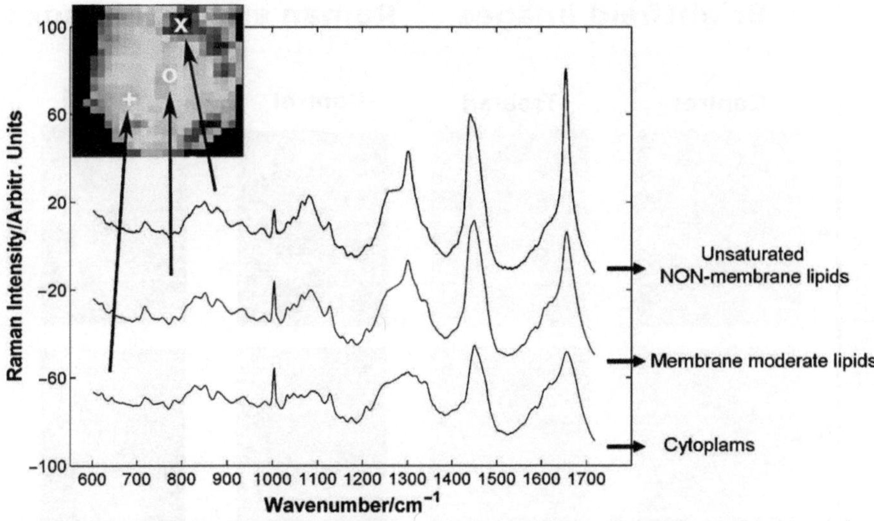

Fig. 6. Raman spectra at various locations within a cell after 6 h exposition to etoposide (reprinted from Zalodek et al. (11) with permission from Wiley Blackwell).

allows spectra acquisition at the sub-cellular organelles level, with features dominated by few biochemical components. Commonly, collection of hundreds or thousands of spectra is required to completely characterise a cellular system. The acquisition time of each spectrum is 1 s. Since a complete Raman spectrum is associated with each pixel, Raman maps can be used to generate chemical images revealing the structure and arrangement of sub-cellular organelles, and can provide insight into cellular biochemical dynamics (Fig. 6). In addition, spectra from different cellular locations can be averaged to generate a spectral biochemical representation of the cell.

16. Raman spectral images corresponding to lipids are built by calculating the ratio between the peak intensity corresponding to lipids (C=C symmetric stretching at 1,670 cm^{-1}) and proteins (phenylalanine breathing mode 1,000 cm^{-1}). Raman spectral images show high accumulation of lipids in apoptotic cells compared to control untreated cells. The concentration of lipids increases particularly at 4–6 h (Fig. 7).

4. Notes

1. For measurements, polystyrene or glass substrates cannot be used due to large amount of background produced that may overwhelm weak Raman signals from the cell background. For measurements, cells are grown on Raman transparent substrates, usually MgF_2, CaF_2, or quartz.

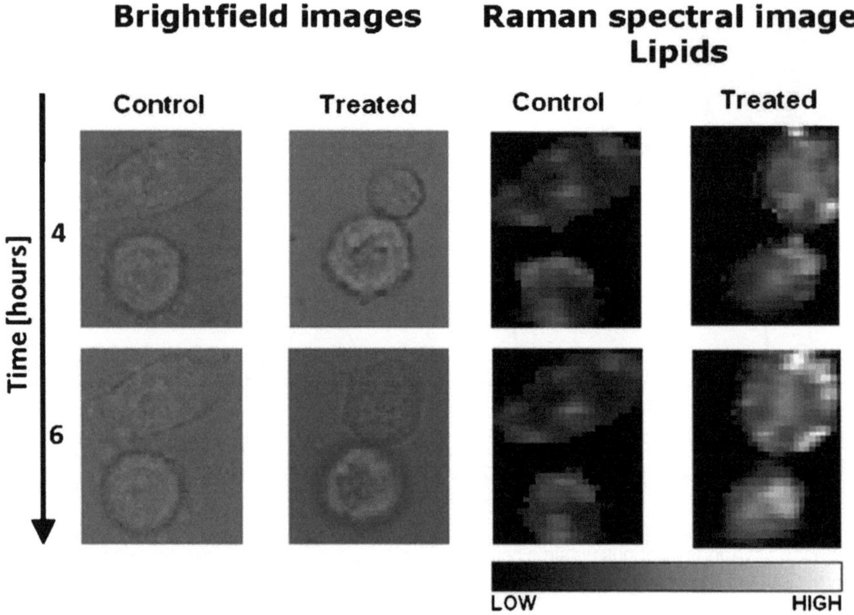

Fig. 7. Time-course optical and label-free Raman spectral images of breast cancer cells untreated and exposed to 300 μM etoposide (reprinted from Zoladek et al. (11) with permission from Wiley Blackwell).

2. Commercial systems are now available (http://www.horiba.com, http://www.renishaw.com, http://las.perkinelmer.com, http://www.thermo.com, http://www.witec.de/en, http://www.deltanu.com, http://www.ramansystems.com, http://www.princetoninstruments.com), but they mostly use upright setup and are designed for wide applications. However, numbers of studies, including experiments on living cells, are reported on home-build setups.

   If required for particular experiments or applications, Raman system could be combined with other imaging methods like fluorescence imaging or Rayleigh scattering imaging.
3. The correct selection of the laser wavelength and power is an important consideration for Raman spectroscopy. The intensity of Raman signal, the autofluorescence of the sample and damages associated with the absorption and heating, has to be considered. As most biomolecules exhibit a minimum absorption of light in the Near Infra Red (NIR) domain, excitation wavelength in this region (usually 785 nm lasers) will minimise the risks of sample heating and/or photochemical interactions.
4. The purpose of filters is to simply cut off non-Raman scattering and stray light, while spectrometer is use to spatially spread the light scattered from sample as a function of wavelength.

5. Sample illumination system and light collection optics refers to a microscope fitted with the appropriate objective and set of lenses to guide the light. In most of published work on using commercially available Raman Microscopy for cells studies, up-right microscopes in conjunction with water immersion objectives are employed dipped in the fluid (PBS or culture medium). The immersion of the objective in the cell culture medium or physiological fluids enables higher Raman signal strength when compared to signal obtained using inverted microscopes, where optical losses occur. These losses are mainly due to the reflections at different interfaces (e.g. cell-chamber window, or coverslips on which cells are cultured). However, the use of the inverted Raman micro-spectrometer offers more flexibility for performing Raman spectral imaging and long-time experiments on living cells, since cells are not disturbed during experiments and issues related to bacterial contamination are avoided.
6. The measurement method is to be chosen according to the application.

References

1. Taylor, R.C., Cullen, S.P., Martin, S.J. (2008) Apoptosis, controlled demolition at the cellular level. *Nat. Rev. Mol. Cell Biol.* **9**, 231–241.
2. Heeres, J.T., Hergenrother, P.J. (2007) Poly(ADP-ribose) makes a date with death. *Curr Opin Chem Biol.* **11**, 644–653.
3. Tait, J. (2008) Imaging of apoptosis. *J. Nucl. Med.* **49**, 1573–1576
4. Balasubramanian, K., Schroit, A.J. (2003) Aminophospholipid asymmetry: a matter of life and death. *Annu. Rev. Physiol.* **65**, 701–734
5. Laserna, J. (2005) An introduction to Raman Spectroscopy: Introduction and basic principles. *spectroscopyNOW.com*
6. Notingher, I., Hench, L.L. (2006) Raman microspectroscopy: a non-invasive toll for studies of individual living cells in vitro. *Exp. Rev. Med. Dev.*, **3**, 215–234
7. Notingher, I., Verrier, S., Haque, S., Polak, J.M.P., Hench, L.L. (2003) Spectroscopic study of human lung epithelial cells (A549) in culture: living cells versus dead cells. *Biopolymers*, **72**, 230–240.
8. Notingher, I., Selvakumaran, J., Hench L.L. (2004) New Detection System for Toxic Agents Based on Continuous Spectroscopic Monitoring of Living Cells. *Biosensors and Bioelectronics* **20**, 780–789
9. Uzunbajakava, N., Lenferink, A., Kraan, Y., Volokhina, E., Vrensen, G., Greve, J., and Otto, C. (2003) Nonresonant confocal raman imaging of DNA and protein distribution in apoptotic cells, *Biophysical Journal* **84**, 3968–3981.
10. Verrier, S., Notingher, I., Polak, J.M., and Hench, L.L. (2004) In situ monitoring of cell death using raman microspectroscopy, *Biopolymers*, **74**, 157–162.
11. Zoladek, A., Pascut, F.C., Patel, P., and Notingher, I., (2010 submitted) Non-invasive time-course imaging of apoptotic cells by Confocal Raman micro-spectroscopy, *J Raman Spectrosc*, DOI 10.1002/jrs.2707 (in press).

Chapter 20

Closed Ampoule Isothermal Microcalorimetry for Continuous Real-Time Detection and Evaluation of Cultured Mammalian Cell Activity and Responses

Olivier Braissant and Alma U. "Dan" Daniels

Abstract

Closed ampoule isothermal microcalorimetry (IMC) is a simple, powerful, nondestructive, and convenient technique that allows continuous, real-time detection and evaluation of cultured cell activity and responses. At a selected set temperature, IMC measures the heat flow between a sample and a heat sink and compares it to the heat-flow between a thermally inactive reference and the heat sink. Since heat flow rates are proportional to the rates of chemical reactions and changes of state, IMC provides a means for dynamically following these processes in any type of specimen – including ones containing cultured cells. The ability of IMC instruments to provide measurements in the microwatt (μJ/s) range allows one to detect and follow the activity (including replication) of low numbers of cells in culture (ca. 10^3–10^5, depending on cell type). Closed ampoule IMC is increasingly being used in medical and environmental sciences. While a closed ampoule imposes limitations, it conversely provides simplicity and excellent control. Also, it is still usually possible with closed ampoules to follow mammalian cell activity and replication for several days. This chapter provides an overview of IMC measurement principles and provides examples of the use of IMC for evaluating cultured human and other mammalian cell activity and responses.

Key words: Microcalorimetry, Cell culture, Cell proliferation rate, Cell growth rate, Cell metabolic rate, Biocompatibility assay, Osteocyte, Fibroblast, Lymphocyte, Virus

1. Introduction

1.1. Overview of Isothermal Microcalorimetry

Isothermal calorimetry studies of biologic processes began in the eighteenth century. The first isothermal calorimeter, devised by Lavoisier and Laplace, used ice to maintain a constant temperature. A living animal was placed on the ice and the amount of ice melted in a given time was measured as an indication of the rate of metabolic heat produced (1). Related modern calorimeters are of course much more sophisticated and sensitive. Isothermal

Martin J. Stoddart (ed.), *Mammalian Cell Viability: Methods and Protocols*, Methods in Molecular Biology, vol. 740, DOI 10.1007/978-1-61779-108-6_20, © Springer Science+Business Media, LLC 2011

*micro*calorimetry (IMC) is defined as the measurement of heat production or consumption at rates within the μW range at essentially constant temperature (2). In modern IMC instruments, phase transition of a substance is of course not used anymore as a monitoring method. However, for IMC instruments which provide a differential scanning calorimetry (DSC) mode, it is still used as a means for producing a known amount of heat to assure that internal calibrations of heat flow are accurate (3). For IMC instruments without a DSC mode, many isothermal reactions can be employed for fundamental calibration (4).

Most modern IMC instruments are *heat conduction* calorimeters in which heat is allowed to flow to a heat sink maintained with extreme accuracy at a constant temperature. A thermoelectric module (comprised of Peltier or Seebeck elements) interposed between the specimen and the heat sink registers any slight temperature difference between the two as an electric signal. A similar measurement can be made between a thermally inactive reference specimen and the heat sink, providing a differential comparison to increase accuracy. In *heat compensation* IMC instruments, a constant temperature is maintained and heat production or consumption is compensated either by using the thermoelectric module described above and/or using an electric heater (5). In both heat conduction and heat compensation IMC instruments the conduction- or compensation-related electric signal is recorded and then reported as heat flow, using the calibration data. IMC instruments also incorporate high-quality amplifiers to produce a final heat flow-related signal, which is easily recorded and manipulated.

Isothermal microcalorimetry has been widely used to evaluate dynamic chemical processes such as cement hydration and to investigate stability of drugs or explosives over time at constant temperature. However, the recent development of multichannel (i.e. multicalorimeter) instruments (see Fig. 1) has made this tool more and more attractive for biological studies (6).

1.2. Recent Use of Isothermal Microcalorimetry for Biological Applications

Isothermal microcalorimetry has been used for many different types of biological studies. Many studies have focused on microbial activities in environmental or medical samples as recently reviewed by Braissant et al. (7) and Wadsö (8). In addition, IMC measurements of microorganism metabolism and growth are increasingly used – e.g. by von Ah et al. (9) and others (10, 11) to test antimicrobial agents. It is possible with IMC to rapidly determine the antibiotic inhibitory concentrations. Also at subinhibitory concentrations one can document differences in antibiotic mechanisms (9). With respect to animal cells and particularly human cell lines microcalorimetry has also proven to be a valuable tool for investigating the efficiency of antiviral agents (12, 13). Similarly, separated organelles such as mitochondria can also be studied using IMC, as demonstrated by An-Min et al. (14).

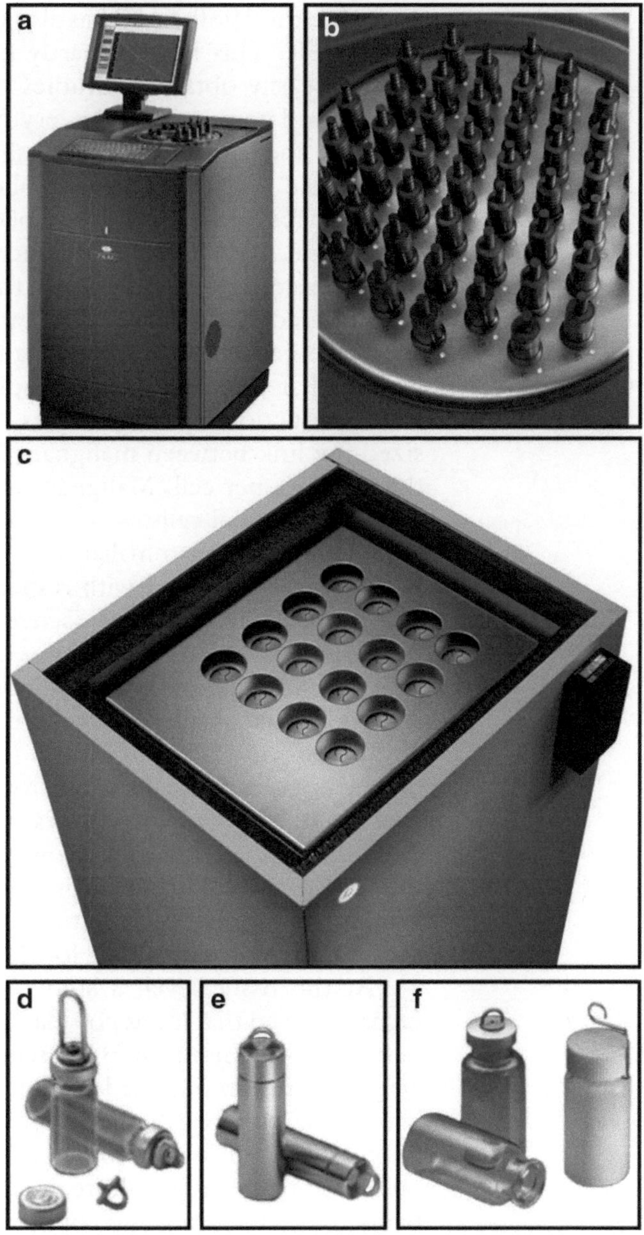

Fig. 1. Examples of commercially available isothermal microcalorimetry (IMC) instruments with multiple microcalorimeters or calorimeters, and ampoules for closed ampoule studies. All instruments and ampoules: Waters/TA, New Castle DE USA (images: courtesy of Waters/TA) (**a**) General view of a TAM III® equipped with various microcalorimeters. (**b**) Top view of a TAM 48® with its array of 48 microcalorimeters (i.e. 48 individual measuring channels). (**c**) Top view of a TAM Air® showing its eight individual differential calorimeters. (**d**) 3- and 4-ml disposable glass ampoules. (**e**) 4-ml stainless steel ampoules. (**f**) 20-ml disposable glass and polyethylene ampoules.

Since the 1980s IMC has also been extensively used to study blood cells. This may be partly because large numbers of red cells are easily obtained. Studies on whole blood and on separated blood components (i.e. erythrocytes, lymphocytes, granulocytes, and platelets) have shown that erythrocytes are responsible for 50% of the total heat output of blood, whereas lymphocytes, granulocytes, and platelets account for 20, 20, and 10%, respectively (15). Differences in heat production rate of blood cells can potentially reveal the presence of several diseases. An elevated heat production rate by erythrocytes has been shown to be indicative of anemia (15). Similarly it was demonstrated that non-Hodgkin's lymphoma resulted in an elevated heat production by lymphocytes (16). The same study emphasized the link between malignancy of tumor cells and heat production rate per cell. Malignant cells produced more metabolic heat than normal cells.

Many other mammalian cells and mammalian cell cultures have been investigated with respect to their production of heat and therefore cell metabolic activity using IMC (17). For example, adipocyte thermogenesis was studied in rat and human cells (18, 19). In vitro studies on adipocytes from obese patients also concluded that an altered cellular metabolic efficiency is an important pathogenic factor in obesity (20). Similarly a study focusing on hepatocytes showed that microcalorimetry could be a valuable tool for investigating carbohydrate and fat metabolism using such cultured cells as a model. However, West and colleagues (21) noted that among mammalian cells there is a discrepancy in heat production rate between cultured cells (e.g. in vitro studies) and cells in vivo (calculated data for the same mammal).

At the tissue level, a study of biopsies of urogenital tract organs showed that it was possible to rapidly discriminate tumorous samples from nontumorous samples using IMC (22). The diagnostic results obtained using IMC in this study were in accord with those obtained by histologic assessment and impulse cytophotometry. Analogous to determining the effects of antibiotics on microorganisms in culture, initial results suggest that IMC can be used for evaluating and comparing effects of antitumor agents on tumor cells in culture (authors' unpublished data).

All the reports of IMC studies performed using mammalian cells emphasize the relative simplicity of IMC methods. Simplicity and other advantages of IMC as a method for studying cultured cell dynamics are described in the next section.

1.3. Advantages of IMC

Sensitivity. With a sensitivity on the order of 0.2 μW, IMC can detect the heat produced by a small number of living cells. Considering that a typical single mammalian cell produces 1–100 pW when active (17), only 2,000–200,000 cells are necessary to produce a detectable signal.

Accuracy. IMC instrument thermostats can be set at any temperature within an instrument's performance range (e.g. 15–300°C) with a maximal error of 0.02°C. Fluctuations around the set point are between 10^{-3} and 10^{-5}°C. During cell growth, the temperature of the microcalorimetric ampoule is maintained within 0.1°C of the set temperature. Baseline drift of IMC instruments is typically ~0.2 μW/24 h. Therefore, for intermediate heat flow ranges (e.g. 20–100 μW) over a few days, heat flow rate measurements can be expected to be accurate within ~1% or less. This provides an excellent means for following metabolism and growth of cultured cells over a period of days since mammalian cell cultures of the size range described above, growing in 4- or 20-ml calorimetry ampoules, have heat production rates of 0.2–50 μW (authors' general experience).

Continuous real-time data. IMC provides a continuous real-time electronic signal proportional to the heat production rate of cells in a microcalorimeter ampoule. The signal must be interpreted carefully, and often requires additional measurements from which to interpret its meaning (e.g. ATP assays, MTT assays). However, insight into dynamically changing processes such as "glycolytic oscillation" (23) can be obtained since the typical heat flow rate sampling frequency is 1 Hz (i.e. 1 data point/s).

Simplicity and passivity. Cells or cell cultures for IMC studies are just placed in a disposable glass ampoule under aseptic conditions, with air or a specific gas mixture in the head space. No fluorescent or radioactive labeling of cells or culture media components is needed. The ampoule is then sealed and placed in one of the measuring channels (microcalorimeters) of an IMC instrument, and heat flow measurements are made as long as there is a heat flow signal of interest (e.g. from hours to days). The signal can be evaluated as it occurs and/or recorded for later evaluation. With both microorganisms and mammalian cell cultures in liquid media, flow-through and flow-stop systems can and have been used. However, they trade control for experimental complexity (24). Sterilization of flow systems is fastidious and time consuming, and raises safety concerns with pathogenic organisms, or cell cultures deliberately infected with viruses. Also, adhesion of microorganisms to the internal surfaces of the flow system potentially compromises interpretation of results (25) unless once wants to study biofilm formation (26). Finally, since heat flow measurements are passive and external, the undisturbed cells and media contained in the sealed ampoule are available for other assays after IMC measurements are finished.

1.4. Limitations of Isothermal Microcalorimetry

Although IMC is a highly sensitive and simple method, it has several limitations that must be taken into account when planning an experiment. The most important limitations are described below.

Equilibration. To obtain the sensitivity and accuracy described above, IMC requires that the sample and (optionally) a thermally inactive reference specimen are precisely at the selected temperature during measurements. In most cases, this requires an overall initial equilibration time of ~1 h, during which data cannot be collected. Flow systems can reduce this time, but introduce the complexities described above. This starting time limitation is not so serious for following the growth of freshly resuspended and diluted mammalian cell cultures. The growth rate is lower than for microorganisms, and there is thus little growth-related heat flow during equilibration. That is, the heat signal remains close to baseline while equilibration is taking place.

Changes over time in the culture environment. The specific IMC research technique described here is closed ampoule IMC of cells in culture. Thus chemical changes such as oxygen depletion and accumulation of metabolic waste products within the ampoule have to be taken into account in interpreting results. Oxygen solubility in water solutions is low. Consequently, in sealed ampoules filled with unstirred liquid medium, dissolved oxygen will be consumed by cells, and the medium will eventually become anoxic. However, the authors have found that a thin-layer liquid- or solid-media cell culture (ca. 1 ml, containing, e.g. osteosarcoma cell lines or periosteal cells) placed in a 20-ml microcalorimetric ampoule with air in the head space allows sufficient oxygen diffusion from the head space into the medium to provide growth conditions similar to those obtained in 6- or 12-well plates. Also, using a so-called nano-calorimeter having a sensitivity of ~20 nW, 150–200 μl of cell culture placed in a smaller 4-ml microcalorimetry ampoule also provides similar conditions (authors' unpublished data).

Nonspecificity of IMC data. Finally it must be noted that the heat flow signal is a nonspecific, net signal related to the sum of all chemical and physical processes taking place in an IMC ampoule. As a consequence, unknown phenomena may produce some of the heat measured, and there may be simultaneous exothermic and endothermic processes taking place, as described by Lewis and Daniels (27). For example, culture medium chemical degradation (often of serum components) can result in a heat flow signal significantly higher than the baseline. Such degradation can last over several days or weeks (Fig. 2a). Similarly, for some experiments such as implant material cellular biocompatibility tests it may be necessary or desirable to sterilize the material specimens using gamma radiation. During IMC measurements, oxidation of free radicals produced in polymeric materials by this type of sterilization can also produce a signal significantly higher than the baseline. For example, see Fig. 2b in Wadsö (8). As illustrated by these two examples, the experimental design must include appropriate controls and reference specimens to avoid systematic errors.

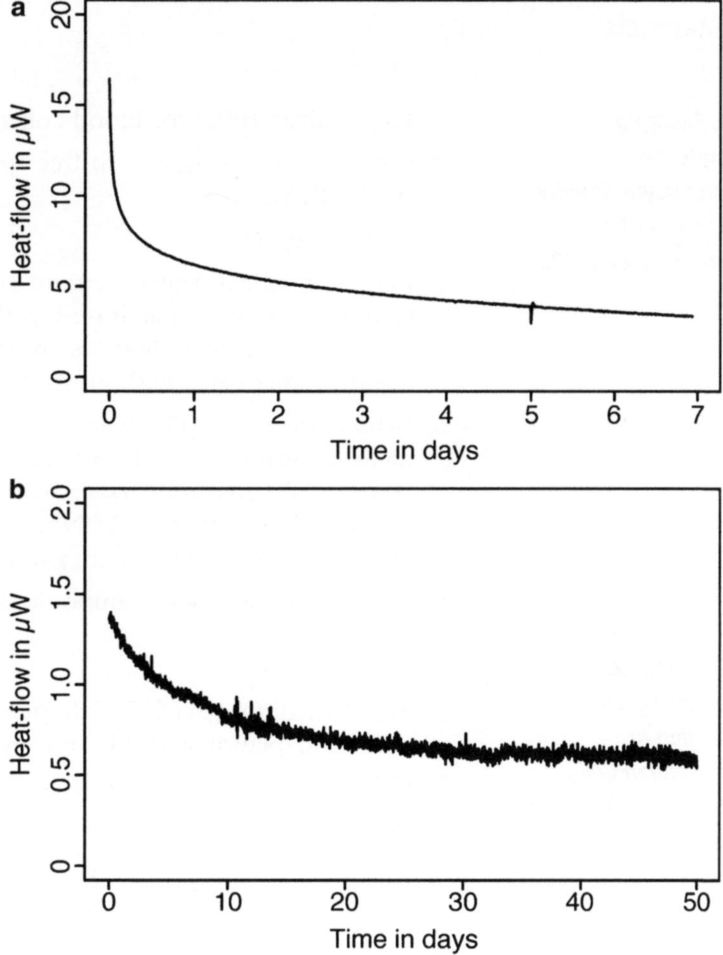

Fig. 2. Examples of extraneous (non cellular) thermal activity from ampoule contents and materials. Heat signals obtained from (**a**) sterile medium containing 10% horse serum and (**b**) sterile microcalorimetric ampoules having undergone gamma irradiation.

1.5. Summary and Expected Future Developments

Sealed ampoule isothermal microcalorimetry (IMC) is a simple to use and highly versatile tool that can accommodate almost any type of cell in (or on) a culture medium, plus a known quantity or concentration of liquid or solid substance (or virus) whose interaction with the cells is of interest. Air or a specified gas mixture can be used in the ampoule headspace. The sensitivity of IMC allows detection and continuous monitoring of cell growth, metabolic rates, or other heat-producing or -consuming activities of low numbers of cells (ca. 10^3–10^5, depending on cell type) in real-time. In addition, IMC is a growing field, and many different experimental techniques and instrument configurations are emerging. The latter include enthalpy arrays (28, 29) and calorimeters on micro-chips (30, 31). It seems likely that isothermal microcalorimetry (IMC) will be used increasingly in biological research and clinical diagnostic settings.

2. Materials

2.1. Example: Monitoring Lymphocyte Activity (Modified After Murigande et al. (32))

1. Heparinized tubes for blood collection.
2. Calcium and magnesium-free phosphate buffered saline (CMF-PBS).
3. Histopaque 1077.
4. M199 or RPMI-1640 medium containing 10% autologous serum or 10% heat inactivated fetal calf serum. No antibiotics are added since lymphocytes are sterilely withdrawn out of a healthy patient (i.e. without bacterial blood infection).
5. IMC Instrument. There are several microcalorimeter instrument companies (e.g. Thermal Hazard Technology, Masscal, Waters/TA Instruments). However, TAM® instruments (Waters/TA Instruments, New Castle DE USA) are used here to illustrate the choices and methodology.
6. 4-ml microcalorimetric ampoule (see Note 1).

2.2. Example: Monitoring Virus Infection in Mammalian Cells (Modified After Heng et al. (13))

1. BHK-21 Cells.
2. MEM medium (pH 7.2–7.4) plus 10% heat inactivated fetal calf serum, penicillin 100 U/ml, and streptomycin 100 μg/ml.
3. Calcium and magnesium-free phosphate buffered saline (CMF-PBS).
4. Trypsinizing solution (0.1% trypsin/0.04% EDTA in Ca^{2+} and Mg^{2+} free phosphate-buffered saline, PBS: 137 mM NaCl, 2.7 mM KCl, 10 mM Na_2HPO_4, and 1.76 mM KH_2PO_4 at pH 7.40).
5. IMC Instrument. There are several microcalorimeter instrument companies (e.g. Thermal Hazard Technology, Masscal, Waters/TA Instruments). However, TAM® instruments (Waters/TA Instruments, New Castle DE USA) are used here to illustrate the choices and methodology.
6. 3-ml glass ampoule.
7. Virus: foot-and-mouth disease virus (FMDV).

2.3. Example: Monitoring Growth of SaOS-2 (Human Sarcoma Osteogenic) Cells as a Materials Biocompatibility Assay (Authors' Unpublished Procedure)

1. SaOs-2 (human sarcoma osteogenic) cells.
2. McCoy's medium (pH 7.2–7.4) containing 15% heat inactivated fetal bovine serum, penicillin 100 U/ml, and streptomycin 100 μg/ml.
3. Hank's balanced saline solution without Ca^{2+} or Mg^{2+}.
4. Trypsinizing solution (0.1% trypsin/0.04% EDTA in Ca^{2+} and Mg^{2+} free phosphate-buffered saline, PBS: 137 mM NaCl, 2.7 mM KCl, 10 mM Na_2HPO_4, and 1.76 mM KH_2PO_4 at pH 7.40).

5. IMC Instrument. There are several microcalorimeter instrument companies (e.g. Thermal Hazard Technology, Masscal, Waters/TA Instruments). However, TAM® instruments (Waters/TA Instruments, New Castle DE USA) are used here to illustrate the choices and methodology.
6. 20-ml microcalorimetric ampoule.

3. Methods

Using IMC to monitor cultured human or other mammalian cell activity.

Since heat flow is produced by many cellular processes, many different types of IMC cultured mammalian cell experiments are possible. Below we first present a methods overview, which is followed by three different specific examples. *However, it should be noted that many other types of cultured human cell studies can be performed using IMC.*

3.1. Methods Overview: Sealed Ampoule IMC Measurements of Heat Flow Related to Cultured Mammalian Cell Processes

Sealed ampoule IMC measurements themselves are simple, innately highly reproducible, and do not require much preparative work. Their usefulness and repeatability depend on whether both cultured cell specimens and materials with which they will interact that are introduced into the IMC ampoules are well-characterized and consistent in both replicate and deliberately varied experiments.

First in IMC, an experimental temperature is selected (e.g. 37°C), and the IMC instrument thermostat is set at that temperature. Typically, if this is a new temperature setting, one must wait a few hours until the heat flow signal from empty channels (calorimeters) returns to baseline.

After the sample (e.g. cells or cell culture – with or without drugs or other additives of interest, medium alone, other control or reference materials) is inserted into a microcalorimetric ampoule, the ampoule is sealed. Each ampoule is then lowered into its own microcalorimeter channel, down to a thermal equilibration position if one is specified by the instrument manufacturer. (Some instruments do not have an equilibration position and ampoules are lowered directly to the measurement position. Similarly some instruments require a reference ampoule filled with a thermally inactive material of same heat capacity and heat conductivity as the sample to be introduced in another microcalorimeter channel in parallel with the specimen ampoule). Lowering the ampoule first to an equilibration position allows the ampoule and its contents to come to a temperature close to the set temperature. This results in less extraneous heat flow when the ampoule is finally lowered to the measurement position. Ampoules are left in the equilibration position, for ca. 15 min (depending on the sample volume, heat capacity, and heat conductivity).

Next, the ampoule is lowered down to the measurement position. In spite of prior equilibration this usually results in a large change in the heat flow signal because of the small (μW) amounts of heat flow being measured. Before starting measurement, one must again wait for relatively complete thermal equilibration, usually 15 min to 2 h or more (depending on the sample volume, heat capacity, and heat conductivity). The necessary amount of time can be determined using ampoules containing material which is not thermally active and has an overall (ampoule plus contents) heat capacity approximately the same as an ampoule containing an active specimen. Finally, heat flow measurements can then take place and be recorded continuously (for hours or days) until the investigator elects to remove an ampoule from the instrument. These steps are summarized in Fig. 3. Also, general measures must be taken in order to avoid specious heat signals (see Note 1).

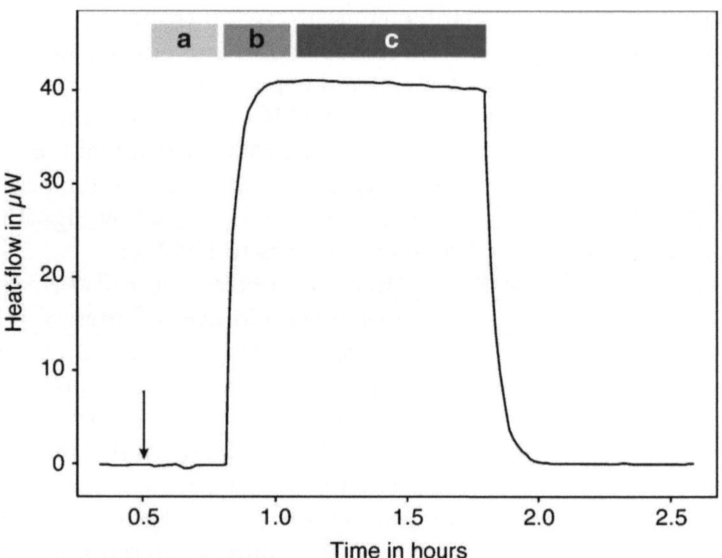

Fig. 3. Example of a typical microcalorimetric measurement for an ampoule containing a specimen whose heat output is slowly declining. The *arrow* indicates the introduction of the sample into the microcalorimeter down to the equilibration position. (**a**) Time interval in equilibration position. (**b**) Time interval for lowering the ampoule into the measuring position and again reaching equilibration. (**c**) Elective time interval in which measurement takes place. At the end of (**c**) the ampoule is removed and the microcalorimeter signal returns to baseline (modified from Monti and Wadsö (33)).

3.2. Example: Monitoring Lymphocyte Activity (Modified After Murigande et al. (32))

Start with an IMC instrument (see Note 2) set and calibrated at the desired temperature (e.g. 37°C).

3.2.1. Isolation of (Human) Peripheral Blood Mononuclear Cells

1. Withdraw ca. 40 ml of blood from a healthy donor in heparinized tubes.
2. Dilute 1 volume of blood sample with 2.5 volumes of PBS.
3. In a sterile conical tube (Falcon) add 10 ml of Histopaque 1077 and 40 ml of the diluted blood.
4. Centrifuge for 30 min at (ca. $365 \times g$).
5. Remove the plasma (i.e. the first liquid layer) and collect the peripheral blood mononuclear cells in a new tube.
6. Wash the cells with PBS and centrifuge at (ca. $400 \times g$) for 10 min. Repeat this operation two times.
7. Resuspend the cells in an appropriate volume of M199 or RPMI-1640 medium augmented with 10% autologous serum or 10% heat inactivated fetal calf serum to obtain a final cell concentration of ca. 2.55×10^5 cells/ml.

3.2.2. IMC Measurements of Lymphocyte Heat Flow

1. Transfer an appropriate volume between 0.5 and 3 ml of the resuspended peripheral blood mononuclear cells into a 4-ml microcalorimetric ampoule. The amount of cells in the ampoule should be between 2.5×10^5 and 7.5×10^5 cells. (If desired, also prepare ampoules with inert reference specimens: usually uninoculated medium of the same volume.)
2. Seal the ampoule (see Note 3).
3. Insert the ampoule in the calorimeter following equilibration procedures described in the Subheading 3.1.
4. Measure activity as long as desired (e.g. 5 days, Fig. 4a).

An example of the type of data obtained with this procedure is shown in Fig. 4a.

3.3. Example: Monitoring Virus Infection in Mammalian Cells (Modified After Heng et al. (13))

Start with an IMC instrument (see Note 2) set and calibrated at the desired temperature (e.g. 37°C) as described in Subheading 3.1

3.3.1. Cell Culture and Preparation

1. BHK-21 Cells are cultured in T-25 culture flasks filled with complete MEM medium at 37°C and under 5% CO_2 atmosphere.
2. Check for the desired degree of cell confluency and absence of signs of contamination.

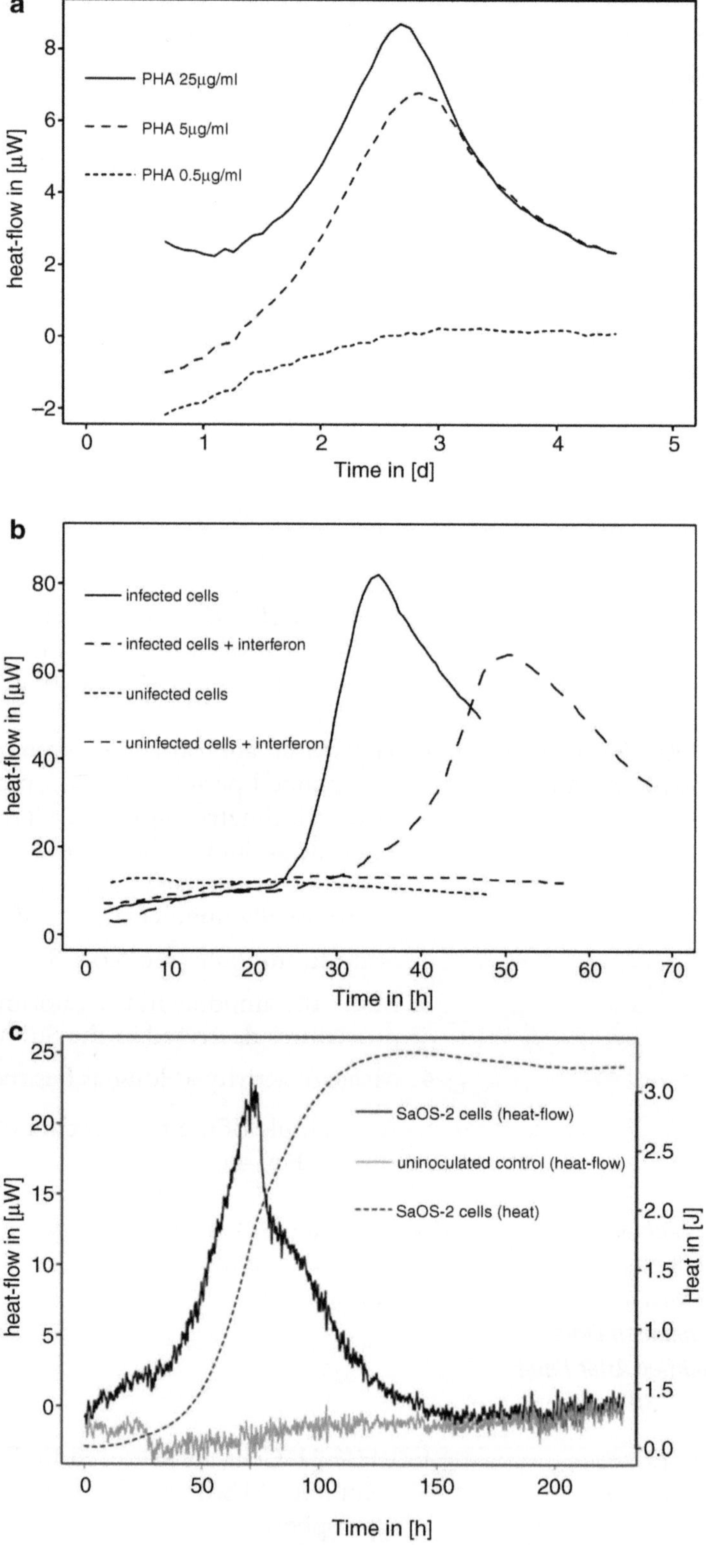
a
PHA 25µg/ml
PHA 5µg/ml
PHA 0.5µg/ml
heat-flow in [µW]
Time in [d]
b
infected cells
infected cells + interferon
unifected cells
uninfected cells + interferon
heat-flow in [µW]
Time in [h]
c
SaOS-2 cells (heat-flow)
uninoculated control (heat-flow)
SaOS-2 cells (heat)
heat-flow in [µW]
Heat in [J]
Time in [h]

3. Remove the spent medium and wash the culture with 1–2 ml of calcium and magnesium-free PBS (CMF-PBS).
4. Add 1 ml of 0.25% trypsin solution in PBS to the culture flask.
5. Gently rotate the flask to cover the monolayer with trypsin.
6. Once the cells are detached and floating, centrifuge and resuspend them in an appropriate volume of MEM medium with added heat-inactivated fetal calf serum, penicillin, and streptomycin.
7. Count the cells with a hemocytometer and adjust the cell concentration to 10^5 cells/ml using prewarmed medium.
8. Transfer 1 ml of the cell suspension to a microcalorimetric 3-ml glass ampoule and allow to grow for 24 h at 37°C and under 5% CO_2 atmosphere.

3.3.2. FDMV: Virus Infection

1. Once ca. 80% confluency has been reached, remove the spent medium.
2. Add an appropriate amount of virus (based on the tissue culture infectious dose – for FDMV virus this is ca. 500,000 PFU or 50 μl of TCID50 virus stock ≈ 10^7 PFU/ml). Incubate for 1 h at 37°C.
3. Add 1 ml of MEM medium along with heat inactivated fetal calf serum, penicillin, and streptomycin.

3.3.3. IMC Measurements

1. Seal the microcalorimetric ampoules (see Note 3) containing the infected cells. If required, prepare an inert reference using uninoculated medium.
2. Insert each specimen ampoule (and reference ampoules if required) into separate microcalorimeter chambers, following equilibration procedures described in the Subheading 3.1.
3. Measure activity as long as desired.

An example of the type of data obtained with this procedure is shown in Fig. 4b.

Fig. 4. Examples of heat flow data obtained following protocols of the type described in Subheading 2. (**a**) Monitoring lymphocyte activity. Figure modified from (32) showing heat flow from lymphocytes (initial amount 2.5×10^5 cells) over a period of days when stimulated with various amounts of phytohemagglutinin (PHA) (original data from (32)). (**b**) Monitoring virus infection in mammalian cells. Example is adapted from Fig. 1 in (13), showing heat flow from BHK-21 cells at 37°C. (a) FMDV infected cells (*plain line*), (b) infected cells with 1.0 μg/ml interferon (*dashed line*), (c) uninfected cells with 1.0 μg/ml interferon (*dashed line*), and (d) uninfected cells without interferon (*dashed line*). (**c**) Monitoring growth of SaOS-2 (human sarcoma osteogenic) cells (authors' data). *Plain dark line*: cells in culture, Plain grey line: inoculated control, *dashed line*: accumulated heat (J) vs. time, base on heat flow curve obtained during growth. Accumulated heat vs. time is analogous to conventional growth curves showing, e.g. the accumulated number of cells.

3.4. Example: Monitoring Growth of SaOS-2 (Human Sarcoma Osteogenic) Cells as a Materials Biocompatibility Assay (Authors' Unpublished Procedure)

Start with an IMC instrument (see Note 2) set and calibrated at the desired temperature (e.g. 37°C) as described in Subheading 3.1.

3.4.1. Cells Culture and Preparation

1. Culture cells in 6-well plates using complete McCoy's medium at 37°C and under 5% CO_2 atmosphere.
2. Remove the spent medium and wash the culture with 0.5–1 ml Hank's balanced saline solution without Ca^{2+} or Mg^{2+}.
3. Add 0.5 ml of trypsinizing solution.
4. Gently rotate flask to cover the monolayer with trypsinizing solution.
5. Upon cell dispersion, centrifuge and resuspend the cells in an appropriate volume of complete McCoy's medium to obtain a final cell concentration of comprised between 1×10^5 and 2×10^5 cells/ml.

3.4.2. IMC Measurements

1. In a previously prepared 20-ml microcalorimetric ampoule (see Note 3) containing a sterilized material specimen to be tested, add 1 ml of cell suspension (1×10^5–2×10^5 cells). Prepare a control ampoule without material. If required, prepare an inert references using uninoculated medium.
2. Insert the specimen ampoule (and a reference ampoule if required) into separate microcalorimeter chambers, following equilibration procedures described in Subheading 3.1.
3. Measure activity as long as desired (e.g. days).

An example of the type of data obtained with this procedure is shown in Fig. 4c.

4. Notes

1. Avoiding specious IMC heat flow signals.

   There are two common sources of specious heat flow signals that can be avoided through experience and good practice.

   First, the ampoule must be completely sealed – i.e. not allow any leakage of vapor from the ampoule during measurement. If leaking occurs, the result is a large endothermic (negative) heat flow signal, which completely overshadows any exothermic heat from the living cells. A screw top which is insufficiently tightened is one source. A less obvious one is insufficient crimping of the metal collar used to seal the rubber septum on glass ampoules. This is a relatively frequent occurrence for new users. Also, injecting substances through the rubber septum with a needle-equipped syringe could result in a septum which leaks. The use of small diameter needles such as 29 or 30 G found on insulin syringes usually solves the problem.

Second, the ampoule must be free of external contamination, which could produce a specious heat signal if the contaminate changes chemically (e.g. oxidizes) or physically (e.g. vaporizes) after the ampoule is placed in a calorimeter chamber. Thus the ampoules should be not only internally but also externally clean, and handled in such a manner as not to contaminate them. With the nanocalorimeters, it is possible to detect a heat flow signal coming from oxidation of oils transferred to the ampoule exterior by handling the ampoules with bare fingers.

2. IMC Instruments

The general principles, advantages and limitations of the IMC method and instruments were described in the main text. Instruments best suited for applied biology (for example, monitoring cultured human or other mammalian cell activity and growth) are those having multiple calorimeters, which allow several specimens to be evaluated simultaneously at a given temperature (6). However, there are several instrument configurations, and the user must select the combination which fits the need. The configuration variables are: number of ampoules which can be evaluated simultaneously, ampoule internal volume, measurement sensitivity, and differential measurement capability. Also, instruments which accommodate more ampoules at once or provide more sensitivity obviously do so at added cost. The authors currently use instruments from the TAM (thermal activity monitor) series produced by Waters/TA (New Castle DE USA). They provide examples below of available configurations for different purposes.

The TAM 48® has 48 separate microcalorimeters, each of which can accept ampoules of either 3 or 4 ml capacity. There is also a TAM III®, which can be equipped with up to 4 so-called nanocalorimeters, which also accept ampoules of 3 or 4 ml capacity. The nanocalorimeters are roughly one order of magnitude more sensitive than the microcalorimeters in the TAM 48®. The nanocalorimeters offer an additional possible advantage. Each nano calorimeter accepts a pair of ampoules, and the resultant heat signal is the differential of the two ampoules. Thus the nanocalorimeter heat signal can be, for example, the difference between an ampoule containing cells and culture medium and a "blank" ampoule containing only culture medium.

Alternatively, if more culture medium or more head space is advantageous, the TAM III® can be equipped instead with up to 12 larger single microcalorimeters, which accept 20 ml ampoules. The TAM III® can also be equipped simultaneously with various combinations of microcalorimeters (e.g. 3 or 4 ml capacity, 20 ml capacity) and nanocalorimeters (3 or 4 ml capacity).

There is also a relatively low-cost TAM Air®, which accepts 20 ml ampoules. Although the TAM Air® instrument's calorimeters are less sensitive (i.e. the measuring range is between 4 μW and several mW), the instrument is configured like the TAM III® nanocalorimeters in that it reports the differential heat signal from pairs of ampoules (up to eight pairs).

Each ampoule-accepting chamber of the instruments of the type described above is a flat-bottomed empty stainless steel tube. (The electronic components which measure heat flow to/from an ampoule are arrayed outside the tube, under and around the sealed bottom.) Except for being at a common temperature, each of an IMC instrument's calorimeter chambers is essentially a separate instrument. Ampoules can be introduced, heat flow measured, and ampoules are removed at any time from any of the chambers with negligible effect (i.e. signal cross-talk) on heat flow measurements in any of the other calorimeter chambers.

3. IMC Ampoules.

The calorimetry instrument manufacturers (e.g. Waters/TA) also supply an array of standardized sealable ampoules for closed ampoule calorimetry, as illustrated in Fig. 1. The glass and polyethylene ampoules shown are intended for single use. The glass ampoules are sealed with an elastomeric septum by crimping an aluminum collar, and this is accomplished with a specially designed crimping tool. If desired, it is possible to inject substances through the septum after sealing, using a needle-equipped syringe. The polyethylene ampoules have a screw-on top. The stainless steel ampoules are intended for repeated use (after careful cleaning to remove any residues). They have screw-on tops with an internal elastomeric seal. The examples below employ glass ampoules of the following capacities: 4 ml (see Subheading 3.2), 3 ml (see Subheading 3.3), and 20 ml (see Subheading 3.4). The glass ampoules are often preferred because they are easily sterilized. The Waters/TA glass ampoules and seals can be autoclaved 20 min at 121°C.

References

1. Lavoisier A. and Laplace P.-S. (1780) Mémoire sur la chaleur, 355 pp. Académie des Sciences, Paris.
2. Wadsö I. (2001) Isothermal Microcalorimetry: Current problems and prospects. Journal of Thermal Analysis and Calorimetry 64, 75–84.
3. Sabbah R., Xu-wu A., Chickos J.S., Planas Leitão M.L., Roux M.V. and Torres L.A. (1999) Reference materials for calorimetry and differential thermal analysis: Thermochimica Acta, 331, 93–204.
4. Wadsö I. and Goldberg R.N. (2001) Standards in isothermal calorimetry. Pure Appl. Chem 73, 1625–1639, calorimetry and differential thermal analysis. Thermochimica Acta 331, 93–204.
5. van Herwaarden S. (2000) Calorimetry measurement. In: Mechanical Variables Measurement (Webster J. G., Ed.), pp. 17.11–17-16. CRC Press, Boca Raton.
6. Wadsö I. (2002) Isothermal microcalorimetry in applied biology. Thermochimica Acta 394, 305–311.
7. Braissant O., Wirz D., Goepfert B. and Daniels A. U. (2009 - in press) Use of isothermal microcalorimetry to monitor microbial activities. FEMS Microbiology Letters.
8. Wadsö I. (2009) Characterization of microbial activity in soil by use of isothermal microcalorimetry. Journal of Thermal Analysis and Calorimetry 95, 843–850.
9. von Ah U., Wirz D. and Daniels A. U. (2009) Isothermal micro calorimetry – a new method for MIC determinations: results for 12 antibiotics and reference strains of E. coli and S. aureus. BMC Microbiology 9, 106.
10. Xi L., Yi L., Jun W., Huigang L. and Q, S. (2002) Microcalorimetric study of Staphylococcus aureus growth affected by selenium compounds. Thermochimica Acta 387, 57–61.
11. Yang L.N., Xu F., Sun X.N., Zhao Z.B. and Song G.C. (2008) Microcalorimetric studies on the action of different cephalosporins. Journal of Thermal Analysis and Calorimetry 93, 417–421.
12. Tan A.-M. and Lu J.-H. (1999) Microcalorimetric study of antiviral effect of drug. J. Biochem. Biophys. Methods 38, 225–228.
13. Heng Z., Congyi Z., Cunxin W., Jibin W., Chaojiang G., Jie L. and Yuwen L. (2005) Microcalorimetric study of virus infection: The effects of hyperthermia and a 1b recombinant homo interferon on the infection process of BHK-21 cells by foot and mouth disease virus. Journal of Thermal Analysis and Calorimetry 79, 45–50. (30)
14. An-Min T., Chang-Li X., Song-Chen Q., Ping K. and Yu G. (1996) Microcalorimetric study of mitochondria from fish liver tissue isolated. J. Biochem. Biophys. Methods 31, 189–193.
15. Monti M. (1987) In vitro thermal studies of blood cells. In: Thermal and Energetic Studies of Cellular Biological Systems (M.A., James., Ed.), Vol. pp. Wright, Bristol.
16. Monti M., Brandt L., Ikomi-Kumm J., Losson H. (1990) Heat production rate in blood lymphocytes as a prognostic factor in non-Hodgkin's lymphoma. European J. Haematology 4, 25–254.
17. Kemp R.B. and Guan Y.H. (1997) Heat flux and the calorimetric-respirometric ratio as measures of catabolic flux in mammalian cells. Thermochimica Acta 300, 199–211.
18. Boettcher H., Nittinger J., Engel S. and Fiirst P. (1991) Thermogenesis of white adipocytes: a novel method allowing long-term microcalorimetric investigations. J. Biochemical and Biophysical Methods 23, 181–187.
19. Boettcher H., Engel S. and Furst P. (1992) Thermogenesis and metabolism of human white adipocytes in gel culture. Clinical Nutrition 11, 53–55. (2) Bataillard P. (1993) Calorimetric sensing in bioanalytical chemistry: principles, applications and trends. Trends in Analytical Chemistry 12, 387–394.
20. Soerbis R., Monti M., Nilsson-Ehle P. and Wadsö I. (1982) Heat production by adipocytes from obese subjects before and after weight reduction. Metabolism 31, 973–979.
21. West G.B., Woodruff W.H. and Brown J.H. (2002) Allometric scaling of metabolic rate

from molecules and mitochondria to cells and mammals. PNAS 99, 2473–2478.

22. Kallerhoff M., Karnebogen M., Singer D., Dettenbaeh A., Gralher U. and Ringert R.-H. (1996) Microcalorimetric measurements carried out on isolated tumorous and nontumorous tissue samples from organs in the urogenital tract in comparison to histological and impulse-cytophotometric investigations. Urological Research 24, 83–91.
23. Lamprecht I. (2003) Calorimetry and thermodynamics of living systems. Thermochimica Acta 405, 1–13.
24. Jespersen N.D. (1982) Biochemical and clinical applications of thermometric and thermal analysis, pp. Elsevier Scientific Publishing Company, Amsterdam.
25. Levin K. (1971) Heat production by leucocytes and thrombocytes measured with a flow microcalorimeter in normal man and during thyroid dysfunction. Clinica Chimica Acta 32, 87–94.
26. von Rege H. and Sand W. (1998) Evaluation of biocide efficacy by microcalorimetric determination of microbial activity in biofilms. Journal of Microbiological Methods 33, 227–235.
27. Lewis G. and Daniels A. U. (2003) Use of Isothermal Heat-Conduction Microcalorimetry (IHCMC) for the Evaluation of Synthetic Biomaterials. J. Biomedical Materials Research 66B, 487–501.
28. Torres F.E., Kuhn P., De Bruyker D., Bell A.G., Wolkin M.V., Peeters E., Williamson J.R., Anderson G.B., Schmitz G.P., Recht M.I., Schweizer S., Scott L.G., Ho J.H., Elrod S.A., Sch ultz P.G., Lerner R.A. and Bruce R.H. (2004) Enthalpy arrays. PNAS 101, 9517–9522.
29. Recht M.I., De Bruyker D., Bell A.G., Wolkin M.V., Peeters E., Anderson G.B., Kolatkar A.R., Bern M.W., Kuhn P., Bruce R.H. and Torres F.E. (2008) Enthalpy array analysis of enzymatic and binding reactions. Analytical Biochemistry 377, 33–39.
30. van Herwaarden S. (2005) Overview of calorimeter chip for various applications. The rmochimica Acta 432, 192–201.
31. Bataillard P. (1993) Calorimetric sensing in bioanalytical chemistry: principles, applications and trends. Trends in Analytical Chemistry 12, 387–394.
32. Murigande C., Regenass S., Wirz D., Daniels A.U. and Tyndall (2009) A Comparison Between (3H)-thymidine Incorporation and Isothermal Microcalorimetry for the Assessment of Antigen-induced Lymphocyte Proliferation. Immunological Investigations 38, 67–75.
33. Monti M. and Wadsö I. (1976) Microcalorimetric Measurements of Heat Production in Human Erythrocytes: IV. Comparison between Different Calorimetric Techniques, Suspension Media, and Preparation Methods. Scandinavian Journal of Clinical & Laboratory Investigation 36(6), 573–580.

Chapter 21

Digital Image Processing of Live/Dead Staining

Pieter Spaepen, Sebastian De Boodt, Jean-Marie Aerts, and Jos Vander Sloten

Abstract

The quantification of live and dead cells in a substrate is often an essential step in cell biology research. A staining protocol that acts differently on live and on dead cells is applied and the number of cells visible is counted using a microscope. Often this counting is done manually or only evaluated qualitatively. If the number of samples to be analyzed is large, counting live and dead cells will become a labor intensive, and in some cases an unreliable, process. The manual procedure also discards potentially relevant information on the cells beyond their live or dead classification. For example, cell size, shape and distribution cannot be measured manually. Thus, developing a software routine to replace the counting process can result in an increase of both efficiency and quality of the data gathering process. Whether or not the time and/or money spent on creating a dedicated computer algorithm is worthwhile, depends on a large number of factors of which some are specific to the samples and some to the technical expertise available. In a large percentage of cases, creating a computer algorithm may be easier than expected. In order for the reader to correctly asses the difficulty level of his/her specific case, an outline on how to tackle the problem is presented within this chapter. The basic concepts of digital imaging, explained in a possible step by step approach, is offered. It is important to be able to estimate the difficulty level for each specific case. Based on a series of questions the potential of creating a computer algorithm can be offset by the costs to be expected.

Key words: Image analyses, Image processing, Live/dead, Quantification

1. Introduction: Basic Concepts of Digital Images

Most information on basic digital imaging can be obtained from a multitude of textbooks (1–5) as well as excellent online information such as http://micro.magnet.fsu.edu/primer/index.html. Although the transition of the analogue to the digital imaging era effectively enabled the quantification of our visual perception, it does not come without drawbacks. During the digitalization process, the continuous analogue signal we perceive is replaced

Martin J. Stoddart (ed.), *Mammalian Cell Viability: Methods and Protocols*, Methods in Molecular Biology, vol. 740, DOI 10.1007/978-1-61779-108-6_21,

by a digital counterpart that only has a distinct number of representational levels. The discretization process is applied on both the intensity and the spatial representation of the image. A good understanding of the impact of both procedures is key in good decision making during the digital image capturing process. In order to display the concepts of the digitalization process, explanatory one dimensional examples of a voltage signal in time are used as well as two dimensional, true image, examples.

1.1. Image Brightness and Bit Depth

A continuous analogue voltage signal as depicted in Fig. 1 (right) contains an infinite number of signal intensities. A digital version of that signal only allows for a fixed number of distinct voltage levels. In order to convert the analogue signal in to its digital representation a decision on the number of levels used has to be made. Because the digital signals are to be stored and converted on a binary processor, the number of levels used to represent the image is chosen to be equal to the number levels possible for several bit depths. Since a bit can either be 0 or 1, the use of two bits corresponds to $2\times2=2^2=4$ different levels. Each bit added multiplies the available discretization levels by two: three bits amounts to $2^2\times2=2^3=8$ different levels and 4 bit results in $2^4=16$ different levels. The effect of the number of discretization levels on the accuracy of the digital representation can be seen in Fig. 1 (left).

The number of bits used is limited by the physics of the analogue to digital converter used, which converts the analogue input signal to the closest digital equivalent. Secondary, using a higher bit

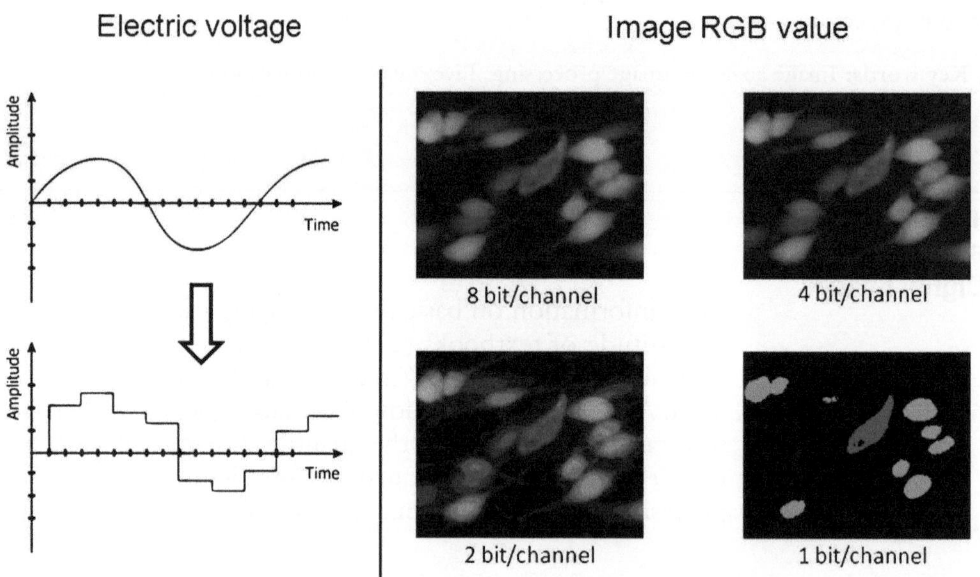

Fig. 1. *Left*: Analogue voltage signal (*top*) and resulting signal after digitalization (*bottom*). *Right*: Effect of bit depth of digital RGB image quality. At 2 bit/channel, the cell texture appears unnatural. At 1 bit/channel cells disappear from the image.

depth results in more memory use and increases computational effort of post processing. For image representation the effect of bit depth can be seen in Fig. 1. An 8-bit digitizer will have 256 gray values to describe each pixel and a 12-bit digitizer will have distinct 4,096 different values. When digitizing a signal, having a high bit depth is not all that counts. Since these bits will be evenly distributed along the whole of the dynamic range of the signal only a portion of them will be in the range of interest. If we would like to measure a voltage signal of ±10 μV with a 12-bit system designed for a ±10 V signal, we will only have 2 bits ($0.010\ \text{V}/10\ \text{V} \times 2^{12} = 4$ levels) describing the signal of interest. Signal conditioning prior to analogue to digital conversion is key in obtaining good results. For image analysis purposes this means setting up exposure such that image contains both very dark and very bright pixels. The human eye perceives a 2% intensity difference in gray levels as just noticeable resulting in roughly 50 different gray levels within the intensity range of a monitor, suggesting a 6 (64 levels) bit representation to be adequate. In order to avoid visually obvious gray level steps after an image has been post processed, 8- or 10-bit resolution is suggested. For digital image analysis, however, even a higher bit depth can be useful. It will allow computer algorithms to detect differences smaller than those detectable by the human eye. Despite the fact that this sounds promising, looking for these small variations can prove to be difficult and may require image processing to boost the differences up to a visible level. As described later, boosting images can result in unwanted emphasis of noise (e.g. Fig. 2, row 3). Noise itself is also often a limiting factor on the usefulness of a high bit depth. Adding additional bits to a noisy signal will not increase the information present in the digital signal. The ratio of intensity range over noise can give a first indication on meaningful bit depth.

Most live/dead staining images are color images and are generally represented by the Red–Green–Blue (RGB) color model, which means that an image is an overlay of a Red, a Green, and a Blue part. Each of these parts is represented by a number of bits. A 24-bit RGB representation will use 8 bits for each color and can represent $256 \times 256 \times 256 = 16{,}777{,}216$ different colors. A 32-bit color space is most often a 24-bit color space with 8 bits for additional non-color data. Higher bit depths (3×12 and 3×14) can be found in current digital cameras and 3×16 bit representation is finding its way into imaging software. Since on screen representation of the image is limited to 24-bit color space, the main advantage of high bit depth has to be found in its ability to retain better image quality after extensive processing.

1.2. Spatial Resolution, Sampling Frequency

Aside from the discretization process to translate an analogue signal into the binary world, a second discretization process is inevitable.

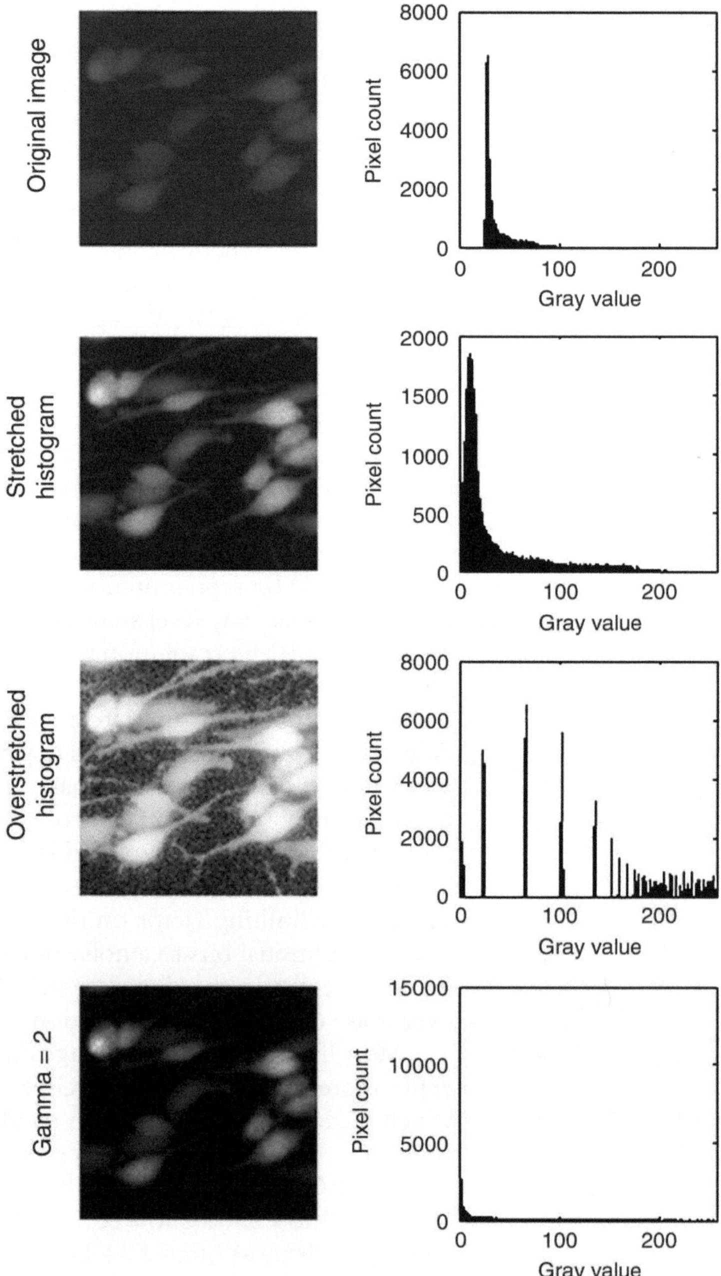

Fig. 2. Manipulation of the gray value histogram and the effect on the image quality.

In image analysis this second discretization is determined by the spatial resolution of the camera system. The effect of undersampling an image can be seen in Fig. 3. Not only does the digitalization process introduce a discretization step, but due to the physics involved, the microscope itself also limits the optical resolution. In order to capture the full potential of the microscope, at least two or preferably three pixels should be available for each

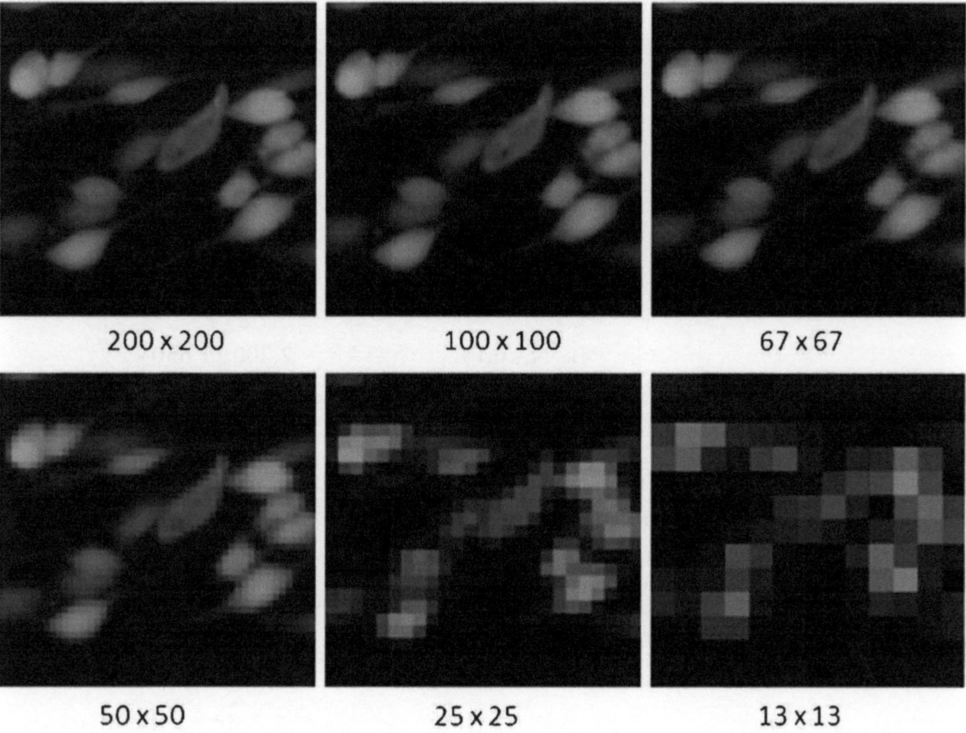

Fig. 3. The effect of spatial resolution on image quality. The effect of undersampling: At 25 × 25 pixels, there are less then 20 pixels/cell and image analysis will be less accurate. At 13 × 13 individual cells become undistinguishable.

resolvable unit. When imaging cells, which are two-dimensional objects, this means that one cell should at least consist of 3 × 3 = 9 pixels. When the images will be used for image analysis, at least 20 pixels per cell is preferable (Fig. 3). Based on the aperture, all magnifications and the physical size of the camera, an optimal number of pixels that make up the camera sensor can be calculated. Adding pixels beyond this optimal number will not increase the amount of information available in the end product. Table 1 gives an overview of commonly used objectives and optimal camera resolutions.

In general, digital cameras used in microscopic imaging are specifically designed for this purpose, but consumer-grade digital cameras can already deliver useful images. Scientific digital cameras, however, have a higher sensitivity, which allows imaging at low light intensities and detects difference in intensity not visible to the human eye. A good indicator of a camera's sensitivity is the quantum efficiency (QE), a measure of the percentage of incident photons that is detected by the camera in a specified spectrum of light. Also important is the camera's signal/ noise ratio (S/N), a measure of the variation of a signal that indicates the confidence with which the magnitude of the signal can be estimated. The effect of S/N level on image quality is illustrated in Fig. 4. Light has an

Table 1
Optimal resolutions for a 2/3″ CDD and 1× magnification of the video coupler

Objective	Optimal resolution
2×, NA 0.06	1,910 × 1,430
2×, NA 0.08	2,590 × 1,940
2×, NA 0.10	3,140 × 2,360
4×, NA 0.14	2,200 × 1,650
4×, NA 0.16	2,590 × 1,940
4×, NA 0.18	2,930 × 2,200
10×, NA 0.30	1,910 × 1,430
10×, NA 0.35	2,200 × 1,650
10×, NA 0.40	2,590 × 1,940

Using a higher resolution will not yield additional information. Objective magnification only, does not include eyepiece magnification

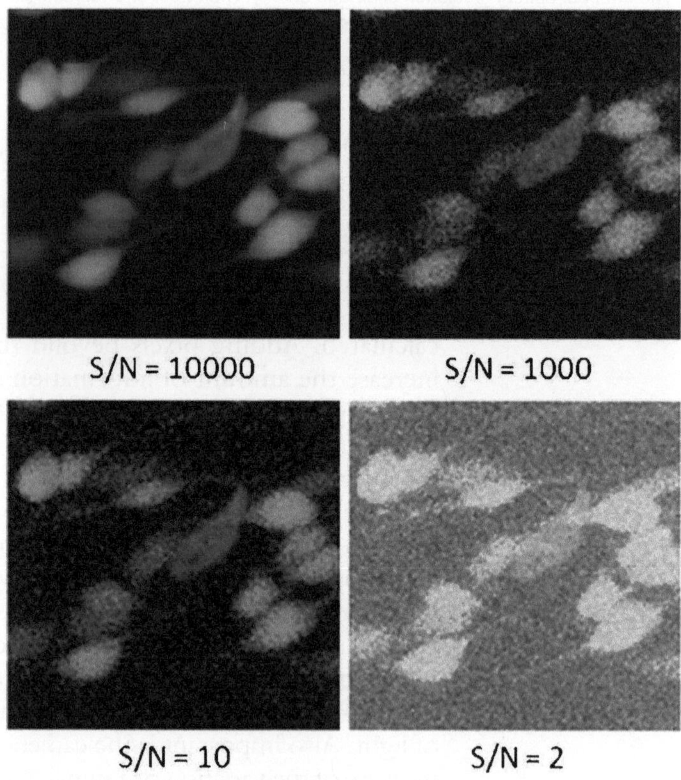

Fig. 4. Images with a decreasing S/N level.

inherent noise component, which cannot be reduced, but noise also derives from a variety of other sources such as the output amplifier. In electronic devices this can often be reduced by lowering the operating temperature. This is why most microscope cameras have an internal cooling. It is the high QE and S/N, and not so much the spatial resolution, which determines the quality and also the price of a microscope camera. Often scientific digital cameras are capable of detecting light in a broader spectrum, unveiling new information, which is not visible to the human eye. Images captured outside of the visible spectrum have to be mapped into the visible spectrum in order to be seen. The invisible part is added to some visible part of the spectrum, potentially masking information already present in this part of the spectrum. In spite of the broader spectrum available, live/dead images typically use the visible part of the spectrum.

Camera software often allows pixel binning, which means that adding the signal of multiple neighboring pixels improves the signal-to-noise ratio and shortens the exposure time. The main disadvantage is the reduced resolution of the final image. When for example applying 2×2 binning, the signal of 4 pixels is added, but the resolution will be four times lower than the sensor resolution. For fluorescent images which generally have low intensity, binning is often used when a high spatial resolution is not needed.

1.3. Image Histograms and Image Evaluation

The main tool to assess whether or not an image processing step will improve the end results is the image histogram (e.g. Fig. 2). The histogram displays the gray or color level, typically between 0 and 255, histogram of an entire image or a selection of an image. During conversion of the image to a binary representation selected intensity values, or value clusters, which should correspond to the cell area will be represented by one and others, which are part of the background by zero. In order for the cells to be retained in the binary data the difference between the cells and the background has to be as big as possible. The more certain intensity levels can be assigned to originate from pixels that are part of the cells, and not the background, the better the binary representation will be. In the image histogram, this will typically be visible as distinguishable regions. The bigger the intensity difference between the cell regions and the background regions, the more feasible it will be to extract them. Displaying the histogram of only a part of an image will help in understanding which intensity levels to look for in the overall image histogram. In some cases, the analysis of the image histogram will yield that there is a significant intensity level between the cells and the background, but no discernible zones in the overall image histogram. In most cases, non uniform lightning conditions will be the cause. This can be verified by looking at histograms from different parts of the background.

1.4. Separating Live and Dead Cells

Because different excitation and emission filters are needed for imaging live and dead cells in one object, the three RGB color channels of the image are taken subsequently and even use a different exposure time depending on the color channel. Depending on the camera software, the images of the live and dead cells are even stored as separate images. For qualitative interpretation these two images can be merged into one, but for image processing it is better to keep them separate (Fig. 5). In an RGB image, only the red channel is used to quantify the dead cells and only the green channel is used for the live cells. Extracting live and dead cells from one image can best be done with two separate image processing sequences.

The first reason is that both have different shape in the image. The calcein, which stains the live cells is present everywhere in the cell's cytoplasm, producing an intense uniform green fluorescence in live cells. Ethidium homodimer-1 which stains the dead cells is fluorescent only when attached to the DNA. This results in only the round nucleus of dead cells staining red. The live cells appear in different morphologies and are more likely to touch, which causes additional challenges for image processing. Dead cells appear as more or less equally sized disks and are generally more clearly separated from one another.

The second reason is that depending on the substrate the cells are attached to, the background in the green and in the red

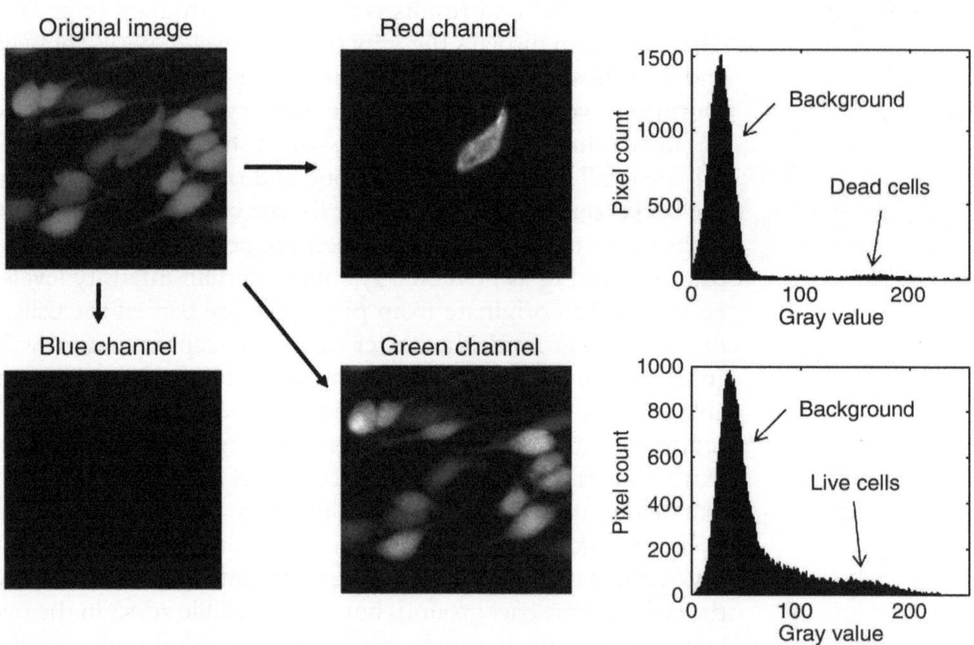

Fig. 5. Separation of live and dead cells by use of *red* and *green* color channel. Live and dead cells often have different morphology and contrast with the background. The latter can be seen in the histograms.

color channel are often very different. This is because many materials are to some extent autofluorescent at the same wavelengths as the calcein or Ethidium homodimer-1. This means that they have similar fluorescent properties as one of these two molecules, which leads to a decreased contrast between the cells and the background and makes segmentation of the cells more difficult. Because this level of auto-fluorescence is different for the red and the green part of the spectrum, the contrast with the background intensity can be totally different for live and dead cells and might require different image processing sequences.

2. Overall Steps for Image Processing

The goal of image processing and image analysis is to allow for (statistical) data processing of features seen in (microscopy) images. Image processing includes all operations on the image and results in new images that have an improved visibility of the cells. During the image analysis that follows, numerical data is extracted from the processed images. From live/dead images, the cell viability [number of living cells/(number of living cells + number of dead cells) × 100%] is the most commonly reported variable. But the same image processing that is required to calculate the viability enables much more elaborate image analysis to determine cell size and morphology or the local distribution of cell viability and cell density. This local information can then be correlated with the local environmental conditions like nutrient or growth factor concentrations.

For verification purposes it is often useful to, in a final step, create a visual overlay of the extracted data on the images. A general guideline for image processing and analysis is:

1. Correct or tone down imperfections, defects, and artifacts.
2. Enhance the cells and suppress nonrelevant information.
3. Convert the image to a binary representation of the cells.
4. Perform measurements on binary image.
5. Overlay extracted information for visual verification.

An example of live/dead image processing using imageJ can be found at the end of the chapter. The steps within this example are referenced to as examples of the principles explained in the next sections. For the original image, see Fig. 6.

2.1. Removing Imperfections

Sometimes only part of the image, which is referred to the region of interest (ROI) is needed for quantification of the live and dead cells. If this is so, it is important to first crop this ROI from your images and to use only this part for image analysis. In this way,

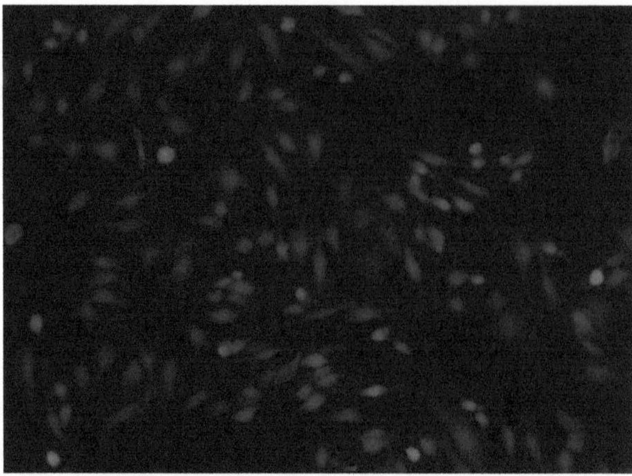

Fig. 6. Example: Original RGB image.

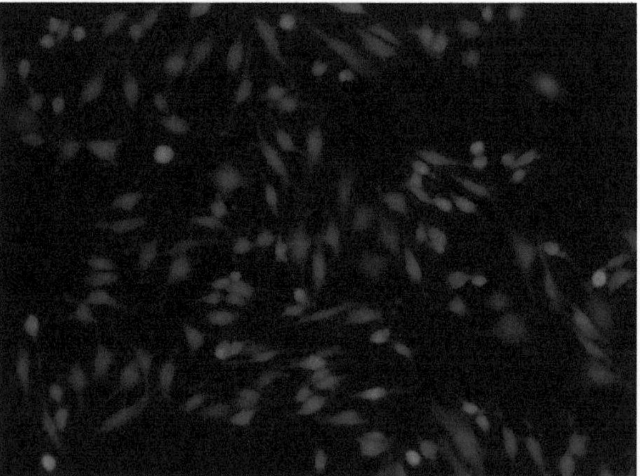

Fig. 7. Example step A: Image → Color → Split channels. Select the *green* channel for live cells.

the effects of inhomogeneous lighting or irregular backgrounds are minimized. Depending on illumination, focus, sample preparation, and equipment used the quality of microscopic images may vary widely. The first step after cropping the ROI form the image is the removal of inhomogeneous lightning, noise, and images defects. Often only one color of a full color image is selected in further calculations [Fig. 5, example step A (see Fig. 7)].

If possible remove nonuniform lighting before image capture: a more homogenous image to start the processing on will have a better chance of successful feature extraction. If this is not possible, capturing an image with the sample removed and subtracting this image from the original images will produce more homogenous lightning. If a prerecorded background is not available the background area of the image itself can be used to

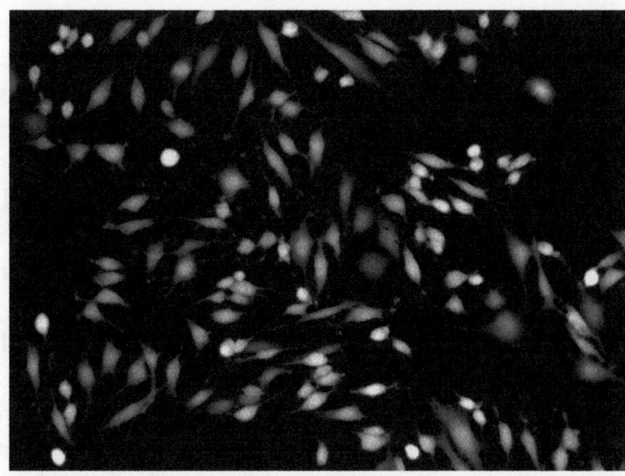

Fig. 8. Example step B: Process→Subtract background: set (rolling ball radius: 100 pixels). Example step C1: Image→Adjust→Brightness/Contrast: set (min = 4, max = 80).

generate a background image [example step B (see Fig. 8)]. Several toolkits can be found on-line that allow the user to select a number of control points of a specific size and use these points to produce and subtract a background. Care has to be taken to allow for difference between different images of set. If lightning conditions are not uniform across all images taken, the background image has to be adapted accordingly.

2.2. Enhancing the Cells and Suppressing Nonrelevant Information

After removing the background the image histogram can be stretched and shifted to make use of the full dynamic range. This is most commonly achieved by adjusting the brightness and contrast (Fig. 2). Care must be taken that the peaks in the histogram that correspond to the cells do not move out of scale as in Fig. 2, row 3. Adjusting the brightness and contrast [example step C1 (see Fig. 8)] does not change the amount of information present in the picture, it merely enhances the visibility of certain details. During the adjustment, the original intensity level is replaced with a new one. Since the number of intensity levels is discrete and finite, changing the intensity level of some pixels will lead to others being be pushed out of bounds or (re)grouped. Changing brightness and contrast is therefore an irreversible process. The same goes for gamma corrections (Fig. 2, row 4). Gamma corrections allow to redistribute the in origin evenly spaced intensity levels. In some cases, more intensity levels are wishful to represent either the brighter or the darker part of an image. If only brightness and contrast are available to stretch and shift the histogram, demanding more intensity levels for the dark or bright part of the image could push the other extreme of the histogram beyond the bounds. This would result in a loss of information.

When using a gamma value of less than 1, dark features in the images become brighter, but the contrast in the very bright and midtone values is lost. For gamma values between 1 and 3 bright features become more dark at the expense of contrast for dark and midtone values.

After these simple image adaptations more advanced filters can be used. In contrast with brightness, contrast and gamma corrections, these filters will not only take into account the intensity values of the individual pixel but also the values of the neighboring pixels. The intensity value of a pixel is replaced by a weighted sum of the intensity of the pixel itself and its surrounding neighbors. The size of the neighborhood can vary between different filters and different filter settings. The average or median and the unsharp mask filter, shown in Fig. 9, are two commonly used filters. Both the average and median filter aim to remove noise (which is visible as individual speckles) by replacing the intensity

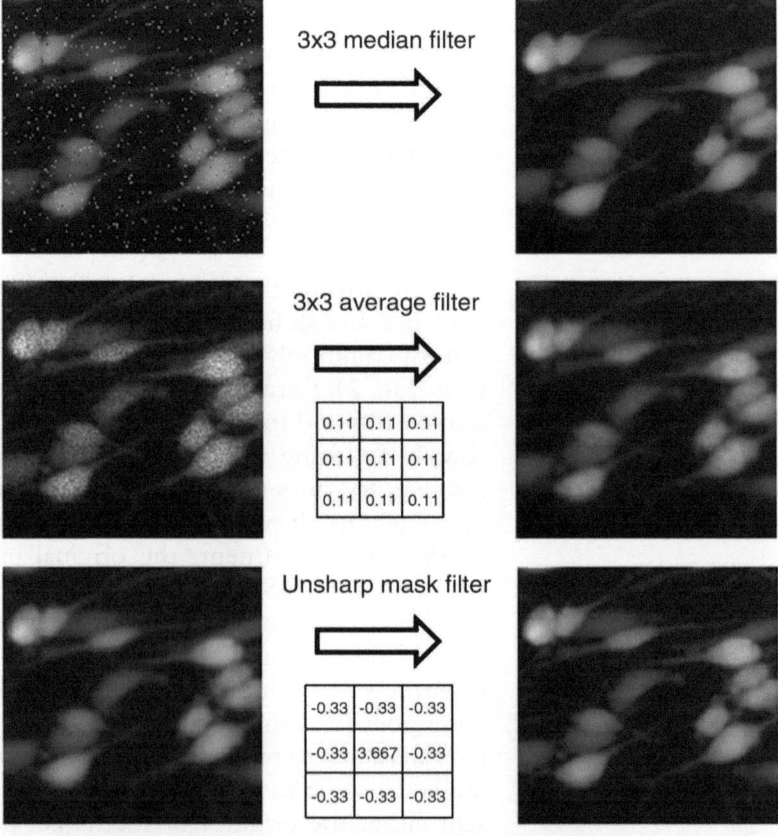

Fig. 9. Examples of commonly used filters. A median filter can be use to remove white and black pixels that originate from broken camera pixels. An average filter can be used to reduce random noise. The unsharp mask filter can be used to restore the sharpness after median or average filtering.

value of a pixel with the average (or median) of the surrounding neighborhood, varying the size over which the average (or median) is taken between 3×3 to 7×7 or even larger [example step C2 (see Fig. 10)]. In some cases, applying the same filter multiple times will further improve the image quality. Removing the dust speckles using an average filter does come at the cost of overall sharpness, the processed image will be free of noise spots but have a more blurry appearance, and neighboring cells are more difficult to distinguish. In such cases an unsharp mask filter may be useful [example step C3 (see Fig. 11)]. In spite of its name, the unsharp mask filter actually highlights details resulting in more crisp and sharpened images. The procedure uses two steps. First a blurred version of the image is created using a Gaussian filter. This filter cancels out all high frequency content (all sharp edges)

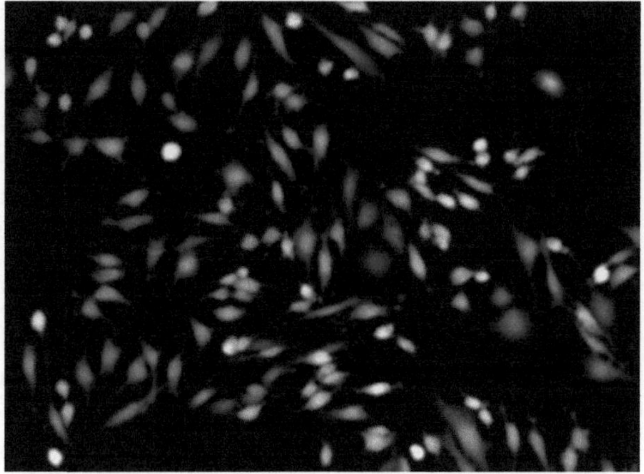

Fig. 10. Example step C2: Process → Filters → Mean → set (radius: 3 pixels).

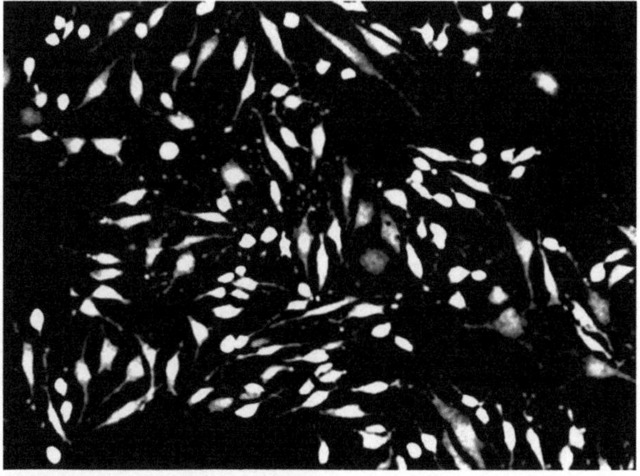

Fig. 11. Example step C3: Process → Filters → Unsharp Mask: set (radius: 6 pixels, mask weight: 0.90).

of the images. The amount of blurring can be controlled by the Gaussian filter settings. A scaled version of the blurred image is than subtracted from the original. This scaling factor provides the user with a second control parameter. The presence of these user-adjustable controls is the main advantage of the unsharp mask filter over other sharpening techniques. Furthermore, because the low frequency content (slow variations) in the images is canceled out, the unsharp mask filter can also be used to remove slow variations in background. The amount of sharpening can be chosen conservatively. Overuse of the unsharp mask filter will blow up noise out of proportion. It is for this reason that sharpening filters such as the unsharp mask filter must be used conservatively.

2.3. Segmenting the Image

After carefully adjusting the image to enhance the intensity of the cells and suppress nonrelevant information, the image can be segmented [example step D (see Fig. 12)]. During this process the intensity value of a pixel is replaced by a 1 or a 0. The most simple case is the one of a grayscale image for which all intensity levels above, or below, a fixed threshold are set to 1 (Fig. 13). A slightly more complex approach requires the setting of two threshold levels: intensity values above the lower threshold and below the upper threshold are set to 1 and all others are set to 0. This method is used when the features of interest have an intermediate intensity value and both darker and brighter backgrounds have to be ignored. This can be useful when certain regions in the background are autofluorescent and have a higher intensity than the cells. The threshold values of the former segmentation methods can be chosen to be fixed for all images to be processed or dependant of some image parameter such as overall brightness. If the background and intensity corrections of the image processing are successful and coherent for all images to be processed, setting fixed threshold,

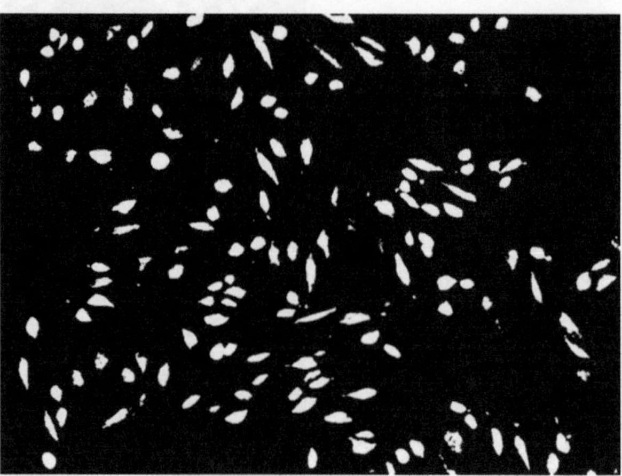

Fig. 12. Example step D: Image→Adjust→Threshold: set (min: 246, max: 255).

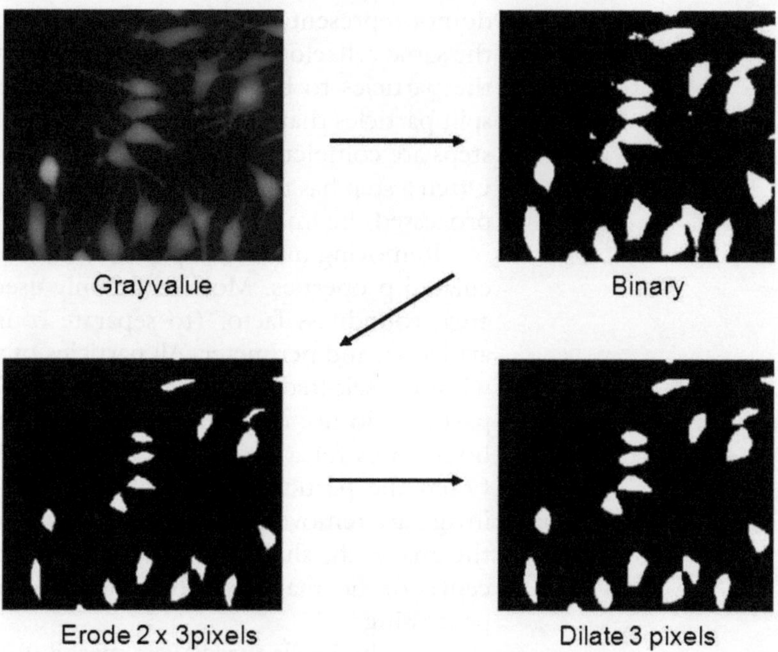

Fig. 13. Gray value image and image after gray value thresholding (*top*). Eroding removes small particles and separates touching cells. Dilatation is performed to restore the size and shape of the cells.

if possible, values will yield a more robust processing method. Creating an intensity histogram of one or more selected features, without background, helps in the selection of proper threshold values. As discussed further on, an additional filtering of the segmented image clears up some of the inevitable thresholding inaccuracies. Some of the more easily removable slip-ups are small speckles of wrongfully categorized pixels and holes inside the cells (as long as there is no connection between the hole and the outside of the cell). Cells that connect to each other prove more difficult to separate and should be avoided. As mentioned earlier this is what makes cell segmentation on live cells more difficult compared to dead cells. Typically, more out of focus cells also yield a bigger challenge in selecting correct threshold values.

In order to assess the quality of the segmentation, several criteria have to be taken into account. After segmentation particles are defined. A particle is a group of pixels that are directly connected to each other. Most software packages offer several definitions of a particle (or hole): (empty) pixels connected along a side (one pixel has four neighbors) or (empty) pixels connected along a side or via a corner (one pixel has eight neighbors). Later in the process, different properties for the particles are calculated, so ideally each cell should already be represented by one individual particle. If not the filtering process has to eliminate particles that do not represent a cell, connect separated particles that represent the same cell, close holes located inside particles, expand or shrink the particles to best mach the complete surface of the cell and split particles that overlay multiple cells. The order in which these steps are completed is not fixed and requires some trial and error. Often a step has to be repeated once more after further steps have processed the image.

Removing unwanted particles can be done based on their calculated properties. Most commonly used parameters are surface area, roundness factor (to separate round cells from elongated artifacts), and perimeter. All particles of which a parameter is not within a selected range are removed. Keep in mind that some particles do not correctly represent a specific cell yet, setting the boundaries for a specific parameter has to be done accordingly. Often the particles that contain pixels at the boundary of the image are removed. Because this could be a cell that is entirely in the image, the shape and size can be different from the cells in the center of the image, which can lead to errors in the further image processing.

Attached cells that have a spread out morphology can have a nonuniform intensity and the darker regions might become black after thresholding, leading to holes inside particles. Filling holes inside particles [example step E1 (see Fig. 14)] will allow for a correct representation of among others the surface area of these particles. This surface area can be used in future filtering steps or could also provide a means to discriminate the results of different experimental conditions. If the outer edge of the cell-particle is only represented by a single line of pixels on one point and pixels are connected via corners also, closing holes can result in unexpected results.

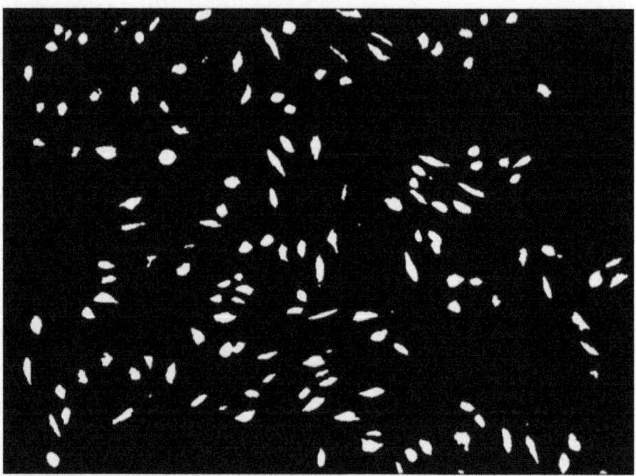

Fig. 14. Example step E1: Process→Binary→Fill holes. Example step E2: Process→Binary→Erode. Example step E2: Process→Binary→Erode. Example step E3: Process→Binary→Dilate.

Eroding particles [example step E2 (see Fig. 14)] can be useful to split a particle that represents two, or more, overlapping cells (Fig. 13). During the eroding procedure the outer pixels of the particle are set to zero. Removing pixels along the perimeter will eventually lead to a split up into two separate cells. This process can lead to unwanted small isolated particles of pixels that used to be part of the original particle. Filtering the image based on particle size will remove such artifacts.

Dilatation of the particles [example step E3 (see Fig. 14)] can be used to better correspond to the actual surface area of the cell. It is also useful to replace several small particles, all representing the same cell, with one larger particle. As soon as two particles make contact, consider the different definitions of contact, they will join to create one single particle. As a side effect of expanding cells, jagged edges become more smooth.

More robust splitting algorithms divide particles representing cells into separate particles. These algorithms refine the rough shrinking process by using additional information like local perimeter curvature. Most software packages provide such procedures, but their result is unpredictable. Avoiding overlapping particles, if possible, remains the better option.

The filtering process [example step F (see Fig. 15)] does not have a specific order of filters to be applied. Often more stringent boundaries for a specific property can be applied once the particles represent the cells more closely. An example of this process could be an image with a large, elongated particle all along the edge of the image in which in the inside is filled with particles

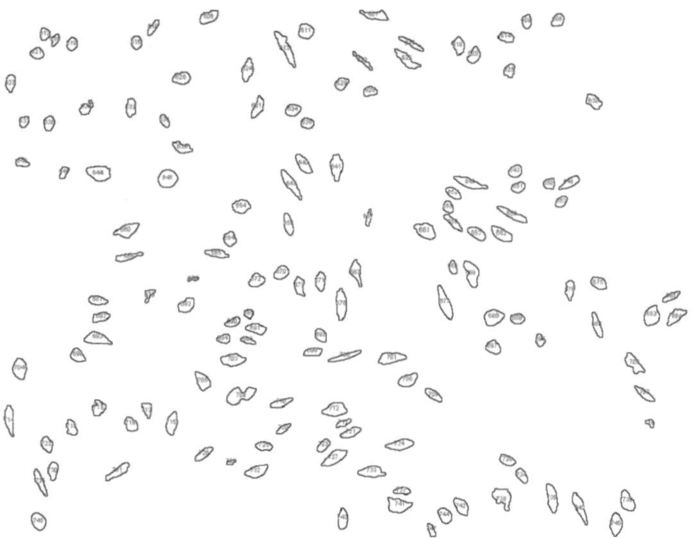

Fig. 15. Image → Lookup Tables → Invert LUT. Example step F: Analyze → Analyze Particles → set (size: 60, show: outlines).

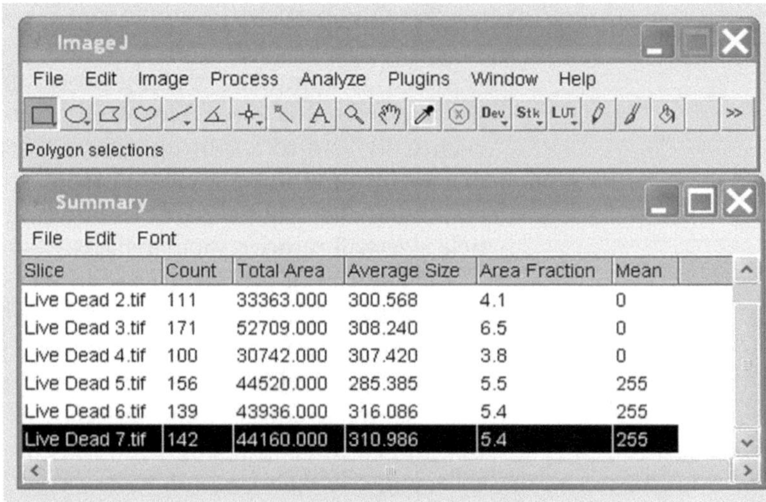

Slice	Count	Total Area	Average Size	Area Fraction	Mean
Live Dead 2.tif	111	33363.000	300.568	4.1	0
Live Dead 3.tif	171	52709.000	308.240	6.5	0
Live Dead 4.tif	100	30742.000	307.420	3.8	0
Live Dead 5.tif	156	44520.000	285.385	5.5	255
Live Dead 6.tif	139	43936.000	316.086	5.4	255
Live Dead 7.tif	142	44160.000	310.986	5.4	255

Fig. 16. Report results, results of six different images are shown.

representing cells. These cell-particles have holes inside and some non cell representing particles, both smaller and bigger, are also present. Thresholding on size will remove the outer, enclosing particle and some of the smaller particles, but because of the holes in the cell-particles, not all small non-cell-particles can be removed. Filling the holes after the removing the large edge particle, followed by a second filtering on size will remove the last unwanted particles

2.4. Performing Measurements

Once the segmentation process is completed successfully the extraction of quantifiable parameters is easy [example step G (see Fig. 16)]. All software packages will present the user with a vast number of available parameters, ranging from the simple number of cells to an extensive representation of the shape and size of each individual cell. Although the simple number of live and dead cells will mostly suffice, it is worthwhile to explore the options: subtle differences between experimental conditions can be picked up.

3. Assessing the Difficulty Level

When analyzing a digital image to detected particular cells against a particular background, no one definitive approach can be made. Existing or commercially available algorithms will therefore often fail to recognize the cells without major adaptation. Creating specific algorithms with specific settings will be necessary for each set of images. A step by step approach with reasonably high success rates is available. Everything starts with the digital image taken. Imaging cells for automatic cell detection demands a more careful approach. With the comments above a well founded decision on the imaging parameters can be made. These parameters should be kept as constant as possible when taking multiple images to make the image analysis more robust. A first assessment of the difficulty level of the project at hand will also direct towards good images.

If using a higher magnification (which produces better quality images, e.g. well separated cells) the downside of having a smaller sampling size can be counteracted by selecting multiple images for each sample. Since the processing will be done automatically, recombining information is merely a matter of good bookkeeping and only requires the additional time of creating the images. To make overview images, use a magnification where one cell is made up of at least 10–20 pixels. As mentioned earlier, the spatial resolution will otherwise be too low to quantify individual cells in the image.

What are the potential benefits?

Do I have to process large numbers of samples for live/dead staining? If so, an automated procedure will improve efficiency. The question on what is a large number, depends largely on the expertise and support available.

Can I benefit from additional information like size or distribution along the image? One of the major benefits of image analysis is the possibility of studying the images in more detail. In the case of live/dead staining, in depth information on e.g. the distribution of live/dead staining can be vital. Time spent on developing counting algorithms can already be justified if additional data allows for a more comprehensive understanding of the experiments.

Is the inter-operator variability of the current methodology too high? Manually assessing viability from live/dead images can be difficult. If the background is of almost the same color as the cells or a lot of cells are present in different focal planes, judgment calls have to be made in order to decide on which cells to count and which not. Training multiple operators to have, and keep over time, the same judgment can be very challenging. Although these images can be among the most difficult to analyze with an automated procedure, the benefits of receptivity can outweigh the costs of algorithm development and the risk of systematic error introduction.

What are the potential costs? In general, the better the images look to the eye, the easier it will be to extract the information needed. In order to assess the potential costs several questions have to be answered.

Are the images expected to be (almost) identical for all experimental conditions and is the background uniform within and between images? The more lightning conditions can remain identical between images the more reliable the counting algorithm will be. For some experimental conditions, it can be challenging to achieve uniform background conditions due to e.g. different substrate thickness. Using some form of background correction can counteract these situations.

Is there a good contrast between the cells and the background? The higher the contrast between cells and background the more easy the separation of the cells from the background will be. Assessing the contrast can be done using a histogram of a cell and its immediate surroundings. A good separation between intensities representing the cell and those originating from the background is preferable.

Are there many overlapping cells? Separating highly overlapping cells is difficult. Some image processing software supply dedicated algorithms, but success is not guaranteed. Avoiding overlapping cells by choosing different magnifications or different concentrations when possible will reduce the time needed for algorithm development substantially.

Are the edges of the cells well defined? The more defined the edges of the cells are the more accurate any measurement, e.g. surface or roundness, on the cells will be. Opting for a higher magnification and compensating the smaller field of view with additional images can improve cell edge definition.

What kind of experience and software can I rely on? The third factor to keep in mind when assessing whether or not to develop an analysis algorithm is the experience and software you can rely on. Large portions of the image analysis up to the point of image segmentation can be done in a large number digital imaging programs targeted at the digital photography enthusiasts. Most of these programs, however, do not offer extensive segmentation algorithms or particle analysis, which can be found in imaging programs targeted at scientific research. Both open-source and commercial applications exist. Splitting up the analysis process with a part on nonscientific and a part on scientifically oriented software can speed up development. Several software options might be available. Without any attempt of being complete, Matlab, Vision assistant, ImageJ (which is free and open source), and Image Pro Plus can be seen as some of the most commonly used software environments for cell image analysis. The background and aim of these programs differ and so does their interface. ImageJ and Image Pro Plus can easily be used to do simple image analysis, without the need for any software development experience whatsoever. The entry level effort is very low and a lot standard procedures are offered. If the specific conditions and are not covered by a standard routine, creating new routines or making substantial changes to existing routine requires the additional effort of learning a, sometimes specific, scripting language. The required effort to acquire the necessary skills can become very large.

Other software, such as Matlab, is used in a much wider field. It targets all engineering problems and not solely image analysis. Additional packages that contain commonly used routines for image analysis can be bought or found and freely downloaded. Adjusting algorithms can be easily done in the Matlab environment, but even solving the simplest of questions will require training in the specific syntax and programming structures of Matlab. If no programming experience is present only very high potential gains can outweigh this effort. The Vision assistant, which is commonly part of academic Labview licenses holds the middle ground. It is targeted toward the development of all types of image analysis routines and provides a large number of standard algorithms. But due to its graphical programming environment, no syntax or programming structure has to be learned. A simple chain graphically representing the sequence of the algorithms to be applied suffices. Which tool is best suited depends largely on the background of the user and the support available.

The more answers to the questions above are positive, the higher the efficiency of introducing automatic counting procedures. Obviously more experienced users will be able to create successful algorithms for more complex cases. The most difficult problems are these of overlapping cells in nonuniform lighting conditions across the image. Even more important than the number of samples to be processed is the potential of the additional information, aside from the simple number of live and dead cells, that can be extracted. The size and shape distribution of the cells or their spatial distribution may hold key elements to discriminate experimental group.

4. Conclusion

Given the proper care in image acquisition, a vast majority of live/dead counting can be done using straightforward image processing techniques. Implementing these algorithms can be done in a variety of software environments of which some are designed to be used by users not trained for software development. A number of the standard image processing software targeted at the market of digital imaging enthusiasts, and even some dedicated engineering software, do not rely on any programming experience of the user whatsoever. Since there is no rule stating that biologists cannot be good at taking and post-processing pictures in day to day life (some argue that biologist take better pictures then engineers), there is no reason they should not be able to implement good live/dead algorithms. Starting within software environments specifically designed to limit the a priori programming experience will help to get familiar with the specific benefits and potential pitfalls of digital imaging processing. The authors hope that by summarizing some of the knowledge on digital imaging and by providing a five-step guideline for the development of successful counting algorithms, the reader will venture (further) on the path of digital image analysis and thus gather higher quality data on their research topic more efficiently.

References

1. **Castleman, K.R.** (1996) ***Digital Image Processing***., Prentice-Hall, Englewood Cliffs, New Jersey, 667 pages
2. **Efford, N.;** ***Digital Image Processing*** (2000) ***A Practical Introduction Using Java***., Pearson Education Ltd., Essex, United Kingdom, 340 pages
3. **Tarrant, J.,** (2007) **Understanding digital cameras: getting the best image from capture to output.,** Elsevier, Amsterdam, Netherlands, 346 pages
4. **Gonzalez, R.C., Woods, R.E.** (2008) **Digital image processing,** Prentice Hall, Upper Saddle River (N.J.), 954 pages
5. **Burger, W., Burge M.J.** (2009) **Principles of digital image processing. 1: Fundamental techniques,** Springer, Londen, United kingdom, 261 pages

INDEX

Martin J. Stoddart (ed.), *Mammalian Cell Viability: Methods and Protocols*, Methods in Molecular Biology, vol. 740,
DOI 10.1007/978-1-61779-108-6, © Springer Science+Business Media, LLC 2011

D

E

F

G

H

I

K

L

M

S

T

V

W

MIX
Papier aus verantwortungsvollen Quellen
Paper from responsible sources
FSC® C105338

If you have any concerns about our products,
you can contact us on
ProductSafety@springernature.com

In case Publisher is established outside the EU,
the EU authorized representative is:
Springer Nature Customer Service Center GmbH
Europaplatz 3, 69115 Heidelberg, Germany

Printed by Libri Plureos GmbH
in Hamburg, Germany